LIFE JOURNEYS

LIFE JOURNEYS

THE WISDOM OF ALL THE AGES

CHRIS THORBY

RETHINK PRESS

First published in Great Britain 2016
by Rethink Press (www.rethinkpress.com)

Cover image © iryna1 – stock.adobe.com

A book of hope for your future,
from a new understanding of the past.

PRAISE

'Very often, history is put into one compartment, and religion into another. What Chris Thorby does in this book—and with great skill—is to begin with a broad and fascinating historical survey, going back even to pre-Christian centuries. He then moves on to a wide-ranging survey of Christianity and its central importance to European societies throughout history. Finally, and very creatively, he explores the relevance of history and religion to how we live our lives today, and to how, in a world facing so many major problems, we can live better and more fulfilling lives in the future.

'This is a book that deserves to be read by believers and sceptics alike. It has certainly encouraged me to rethink my own scepticism.'

Peter Udell read history at Oxford. He then worked in the BBC World Service as a scriptwriter, producer and Service Head, before becoming Controller of the European and Overseas Services.

'This is a very brave, bold and ambitious book. Chris Thorby traces the great sweep of the development of western civilization, and Christianity's role within it, in order to focus on the contemporary challenges we face such as the population explosion, energy shortages, climate change and atmospheric pollution. Fundamental themes, like Jesus's emphasis on the Kingdom of God are highlighted, always with an eye on their relevance to current concerns. He looks, optimistically, to a reconnection between science and religion as helping to make possible the transformation we need. This is a book full of hope. The great message is that we can change things. For Churches it will certainly need a new Reformation if it is to be acted on.'

Rev. Dr. Martin Camroux is Chair of Free to Believe, the United Reformed Church's Liberal Network. He has recently written Ecumenism in Decline: How the United Reformed Church Failed to Break the Mold, and was Times Preacher of the Year.

*Dedicated to Annie
my wife and loving companion
on our own journeys.*

CONTENTS

INTRODUCTION
How To Use This Book

This is a book which asks you to learn a bit more about the journey you're taking through your life. It offers you some ideas about how to take control of this journey and maximise the joys and value of living. Slowly, with a progressing system, it explains how to do this, with one theme leading to another, before arriving at the good news. This is a series of key points for enriching the way we spend our remaining time. So, sadly, your reading experience will not be like dipping into a novel, where you can turn to the last page and see whodunit. You will instead get to the climax only through growing in understanding as each section builds up the knowledge of how the world has grown around you.

Let's start by thinking about this concept of taking a journey through our lives—a familiar, pictorial way of describing the personal sequence of days, months and years which add up to form our own story. As the book unfolds, it asks us to start thinking about this, our personal history. Are we just swirled around in it like pebbles being washed along on a beach by the power of huge waves? Do all the circumstances around us just sweep us along, without us being able to do much about them?

Are we, in fact, trapped by these huge pressures and unable to escape them? What's the consequence of focussing on that view? Is it one of accepting the flow of the tide, and just taking from it the comforts we can find, as they appear along the way? Perhaps comforts of money, security, help and protection from others? Or are there other ways in which we can free ourselves from the trap and start to enrich our lives; giving them more control, growing a new sense of purpose, and finding more depth and enjoyment in what we're doing? Hold on to this the concept of pebbles being swirled along the beach—it will keep popping up to introduce new ideas in the book as they occur.

In the countries of the Western world, we have all grown up in an enormous dominating civilisation, and if we start to look at this anew, we are led to ask three questions:

1. Where Have We Come From?

How have people who've gone before you, right from the dawn of recorded history, searched for key guidance, values and truths, and passed them on to those who followed them?

We'll look at the questions that our predecessors have asked; at the doubts and fears that troubled people; and the ways they found of overcoming them. We'll study how they discerned what was true/useful, and what they learned to be false and discardable. And we'll discover a continuous chain of information being passed down through every generation. We'll also find out how it is that in terms of human emotions and attitudes, past challenges are surprisingly similar to present ones, and that many of *their* solutions remain highly relevant to *us* today. Behind this emerges the understanding that our past has moulded our present. So it is not to be ignored or disregarded. Chapter One will fill out the details of where we have come from.

2. Where Are We Now?

Emerging from this vast environment, like huge waves sweeping along the beach, what are the pressures, hopes and fears which surround us today? Here the story will lead us through some issues which are starting to change the shape of our whole civilisation.

Our hugely increasing world population is one of these, with the result that it is no longer possible for local areas to feed themselves, and not even for whole regional states to do that either. We have to rely on our links with the global world. This in turn produces problems of being neighbours to peoples far away with different cultures and expectations. The transport involved, stretches the need for more of earth's resources of water, oil, and mined minerals. This emerging new factor emphasises the pressing need for people to practice better stewardship of their environment; whether preserving earth's precious resources, or the well-being of our communities, or our own personal health and lifestyles. It ties in with the growing problem of resolving the entirely new challenges of a global world, with its clashes between short and long term gain, and differing cultural/ethnic ambitions. The complexity of modern society has formed a new spirit of individualism about working for oneself, rather than for a team or community. What the political philosopher Thomas Hobbes foresaw in the 17th century, has become a fact of life today.

Additionally, there is a widespread sense of mistrust in authority, an economic storm threatening, and a fading away of faith in the power of religion to find answers to current questions.

All of these form a downside of our present life journeys. But they need to be set beside the remarkable achievements which later pages will spell out. They are both now changing the shape of the world we live in, and their challenges demand responses; not tomorrow—but today.

3. WHERE ARE WE GOING?

The third question explores today's problems in more detail. In an extended section covering Chapters Five, Six and Seven, we will examine how we can use past experience in handling change to guide us today. This section will look to the future to examine the convergence of scientific knowledge with newly emerging themes of faith and morals. Does this new development have the potential to set people free from many of the problems of our present generation? Lastly, we will ask how this suggests ways in which we too can share in maximising lifestyles and finding targets for our own journeys into the future.

————————

CHAPTER ONE
WHERE HAVE WE COME FROM?

Sometimes events occur which jolt us from present pre-occupations to realise that so much of what has gone before, explains how things are today. This came to the front of my mind a few years back when my wife and I sat in a mobile home during a windy night on a Suffolk cliff top. As the gale roared about us, we turned on the radio to seek the comfort of voices in less threatening places. Then there was a news flash. The storm had breached the Suffolk coast and swept a wall of shingle down the beach towards the cliff where we were perched. As soon as light dawned, curiosity overcame common sense and I walked over to the scene of the emergency. Yes, the tide had burst through the shingle bank, and flooded the fields behind for as far as I could see.

But even more interestingly, it had gouged out a great gash in the beach, leaving today's fishing boats perched precariously over a gap… where the foundations of yesterday's buildings had suddenly re-appeared. There was the evidence of the Victorian lifeboat station, long since buried under present changes. Then, beneath that, I realised I could see the remains of earlier buildings at Dunwich—reminding me that I was standing on the site of a great medieval city, one of the largest in Europe—long since buried under the changes in later generations.

So where have we come from? Does it matter? Yes, I should just think it does! For it explains who we are, why we think as we do, and gives us the equipment to face the challenges of our own times.

Stories and analogies help us to decode this, and the images from the beach will appear and expand later in the chapters which follow. But first we need to look at our roots, starting from the very dawn of recorded history.

START WIDE

There is little we know of the span of human existence in the 800,000 years of its evolution from earlier two-legged life forms. Just about 5,000 years—less than 1% of that time—has been recorded in any surviving way. So that is the area of history, everything before that is conjecture about paleoanthropology and other ologies.

During every century of recorded history, there have been instances of human knowledge being collected, passed down, and becoming the basis of wisdom and knowledge which fed the growth of later civilisations. In their turn most of these have risen and then waned, sometimes with larger systems overtaking them. So a really comprehensive study of our origins would have to delve into this whole area; into civilisations of every main geographical area—Europe, the Middle East, Egypt, China, India, Europe, the Americas—and more. But have no fear—that is one Journey we won't be making here. Instead I suggest we draw our examples from the roots of modern Western civilisation—now loosely described as the First World—and from the Classical and Abrahamic cultures from which it evolved. They are taken from the records of religious rather than scientific information because that was where the original store of ancient knowledge mainly resided.

How Knowledge Began to be Recorded

When human experience and attitudes to living started to be recorded, all the wisdom and practical know-how was passed down to following generations within a framework that came to be carefully preserved. Because Stone Age man could no more describe the fundamental questions of life and death than us, collections of basic beliefs and good practice began to be drawn up to guide communities. Practices that were good for the well-being of individuals and communities became identified as truths, and those which caused suffering or disasters became falsehoods. In the stark and simple lives of people at the dawn of history, there was no other way of finding the 'truth' about where we came from, and where we were going. This work of expressing these truths was done by what we would today call religious leaders, though always closely linked with those who wielded practical power.

An important factor in this early growth of human wisdom, was that it was tested by time. That was its pedigree. As we shall later see, thousands of years further on, a new source of human knowledge emerged with the Renaissance in Europe, the great explorations, and the rise of pragmatic investigation and measurement. Two alternative, competing forms of human knowledge stood side by side. We will look at these now, using the rough and ready labels of 'Science' and 'Religion'.

How Information was Passed Down Through the Generations

Before we delve into the gradual and fluctuating growth of our Western civilisation it's worth studying how knowledge gets transmitted from age to age. It's a pattern that occurs in every generation, and which is continuing now.

There are four steps:

1. RECEIVING: Our means of becoming aware of our inherited foundations take many forms. In some cases, they have been bound into the laws of states, and in others into folk lore and traditions. As in the earliest societies there remains a strong religious legacy, both in scriptures and in church traditions and liturgy. The large body of Hebrew and Christian writings, for instance, have profoundly influenced the structures, laws and moral values of most European states—so we shall start from these. Further glimpses of our foundations also appear in the language we use. When we call someone a 'moron' we're linking back to the language of classical Greece nearly 3,000 years ago, to a word meaning dull or stupid—the perceived attributes of the local population of the town of Moron, on the north east fringes of the empire. When we describe something obvious as 'plain as a pikestaff', we link back to military circles of the late 17th century, when the poor old pikeman could hardly

conceal his weapon, as it was sixteen feet long. The past surrounds us with its vocabulary and values, and our starting point for modern life's journeys inevitably spring from them.

2. REJECTION: The second step gets taken because many of the images which were useful for past generations no longer fit our own current situations. As a young schoolboy in the Second World War, I remember singing *Onward Christian Soldiers*, with gusto. A good tune—but look at the words! Written in 1864, it forms the inspiration of three following generations, in images of an army of Christians, marching onwards —where to? To war. Who was in charge of this army? Jesus—that humble charismatic from 2,000 years ago had become, 'Christ the royal master, (leading) against the foe'. References to Satan's host (the other side) and Hell's quivering foundations abound, and the hymn comes to its climax with the happy followers giving Glory Laud and Honour to Christ the King. This piously held view, which guided millions of people in the early years of the 20th century, is no longer a foundation stone for Christians to build on today. It is not to be ridiculed though, and may come to be held again, but for the present, it needs to be put to one side, and other images found to link the wisdom of the ages to our problems of the present. Which leads us to the third step.

3. ADJUSTING/ADDING: This sometimes takes the form of adding new perceived wisdom, or re-interpreting what was already there. This process can be seen at work right from the earliest Christian generations, by carefully reading New Testament writings—particularly in the Gospels and early letters of Paul. Here a picture emerges of the original earthly born Jesus living in Palestine[1] from about 5BCE to 33CE. An itinerant preacher and prophet—one among a number of others in the expectant atmosphere of the 1st century—with the Israelite[2] population seeking release from the punishing regime of the occupying Roman military government. In these volatile times Jesus emerges as a movement leader, calling people to give up earthly pleasures to follow him. Like John the Baptist just before him, he looks to an act of heavenly deliverance from the present, but adds one distinctive element to his call, which can be achieved by ordinary people directly through his agency.

Using a phrase obscure to our modern ears, he calls his people to join in building a 'Kingdom of God'. A state of living where all existence was guided by Yahweh's will, and not by the intentions of humans.[3] This phrase, which occurs over thirty times in the accounts of his preaching, implied two requirements on followers. One was to join up *now*, and the other was to give up all interest in earthy status, family ties or possessions. The early writings focused on this social and spiritual mission, and urged Christian followers to spread the news quickly, for the end of the age was in sight, and a final stage of history was about to begin, where Christ would return, the kingdom of god would come and rule supreme—not through might of arms, but through releasing God's power in men's hearts.

So these were the understandings passed on the next generation, but were they accepted for the new generation to live by? Not really. Adaptations had to be made as the whole spirit of the age altered. The Roman yoke tightened, local genocides took place, and eventually in the mid-sixties of the 1st century a rebellion followed, which in turn was savagely crushed once again. Worse was to follow, with the growth of a state Roman control system in which the commander in chief of the military, became Emperor, and the Emperor became God.

This left the Christians, persecuted both by the Roman powerful, and the Jewish sects from which they were splitting, escaping from Israel and spreading across the whole face of the known world. New Christians in the Levant, in Asia Minor, in Greece, in Rome, in Egypt were people who had never known or seen the earthly Jesus. They needed him to have more pulling power than the rival Roman Godhead. The new writings springing up around this Diaspora (the gospels) all changed and extended the previous generation's message. Jesus, the man suffused on earth by the spirit of God, became the cosmic Christ, present from the start of history, and able to return in glory to judge people at the imminent end of the age. So John's gospel, written at the end of the 1st century, or even a little later, starts with a new creation story, to set beside the older Jewish one.

'In the beginning was the word... through him all things were made. In him was life, and that life was the light of men.' The 'word' was Jesus. One in the eye for the Roman emperor. There are similar stage-three steps to be seen in virtually every generation since—even 'Onward Christian Soldiers' and beyond.

But a note of caution: Adjusting/Adding is not to be seen as a deceit, nor a total rejection of the past, which is itself based on multiple images. It can instead be viewed as an inevitable coming to terms with a new present. Furthermore, there are those who can argue that it shows a further stage of the Creator's revelation, made possible as man's understanding and sensitivity grew. Take your choice, but don't miss the fact that it happened then, through all the ages, continues to happen today.

This transmission of beliefs through history has needed one further step—one which has occurred in every subsequent generation.

4. PASSING INFORMATION ON: This a key element. No civilisation can continue to grow without the main body of acquired wisdom being passed on to the next era. History shows that positive steps have to be taken to make this happen. It's not enough to make records, write or create art—action has to be taken to pass them on. There are three ways in which this has been done in the past.

The first is the onward transmission of specialised knowledge. In the medieval era this might have been the design skills of the castle or cathedral builders. Or the art of illuminating the very few scrolls and manuscripts that contained key wisdom. In later ages a huge mass of agricultural, industrial and scientific knowledge had to be passed down, and in these and many other ways, great institutions emerged to set up training systems to pass on their specialisms. A second factor to note was the missionary tendency as societies became more advanced, and more open. So wisdom was not just passed on by specialists to specialists, but made available to others to grow and develop their own skills. Lastly, all civilisations have shown great care in passing down family or tribal values. Every age wants its children to be shown what the past has learned about healthy and good living. Not to do this is to start a downward plunge in social cohesion, and when the young have lost their way in the past, huge problems have followed.

Later chapters will examine the relevance of this for today's society, and for our own journeys. For now, let's just observe that a neighbourhood, a whole society and, for Christians, the church, moves into crisis without feeding the minds of its young with survival wisdom. Such a crisis would be terminal for the civilisation.

Having seen how human knowledge grows by being passed down through the generations, we now turn to our own Western culture, and briefly journey through the stages which have produced the world we live it today. That will then form the base for examining the present and future of our own life plans and hopes.

NOTES TO CHAPTER ONE

1. *Palestine*. For over three thousand years the narrow land corridor between the eastern end of the Mediterranean Sea and the great desert sweeping away towards the Persian Gulf was a link between the peoples of the African and European continents. Through this corridor migrations, armies, and trade has constantly flowed, and its ownership and local population change with it. Parts of Lebanon, Jordan, Syria and modern Israel span this area today, and to save listing all the changes in old testament times, I use the label 'Palestine' in the text.

2. *Israelites*. To follow the development of one group is also to find a succession of different tribal names for groups of inhabitants separated by linguistic and cultural variation. To save complications, I use 'Israelites' in the text (see *Note 1* above).

3 *Yahweh*. The tribal God of the Stone Age kingdoms of Judah and Israel.

CHAPTER TWO

THE RISE OF
WESTERN CIVILISATION

HOW IT HAS EVOLVED

The roots of today's western world can be traced back
to the dawn of recorded history. To understand our
present, we now have to look back into this past, to
sense its enormity and significance for all of us now.

Our present civilisation is the greatest of all those
that have preceded it. In its geographical spread, its
hugely growing population and its mass of advanced
knowledge, it is the pre-eminent world culture, and
one which has absorbed a number of others as it grew.
In conquering problems of travel, exploring space,
extending human life, and a whole range of technical

developments, the West has vast resources. So great are the human consequences, that a whole new series of lifestyle and stewardship problems are quickly being generated, leaving generations following on from ours to find some fundamental answers.

Its emergence spans a period of about three thousand years—not ages of continuous steady growth, but eras of fits and starts, with a huge acceleration in the last six centuries, one that seems to be quickening on a logarithmic scale in the last 2,000 years. During this whole period, the basis of human knowledge began to be founded in the areas of religion, but later transferred to those of science. These two sources will be explored in more detail later.

Its Roots in the Ancient Greek World

Much early evidence can be retrieved from the recorded history of classical Greece and Rome, and from the early Jewish scriptures, which later formed the framework for Christianity. From them we see knowledge being passed from Egyptian civilisation and passed across the Mediterranean Sea by trade links to the Greek world. There a unique small civilisation began to grow in the 5th century BCE, recorded by contemporary historians, like Herodotus and others, in the next six centuries. Art, architecture and learning flourished, as did a cult of human fitness and competitiveness. Philosophy became established as a means of testing and refining human learning, and of passing it on. Although later centuries have taken a rather idealised romantic view of the Greek world, its contribution to what followed has been immense, through language, mathematics, sciences and a whole range of the arts. Many of the Greek stories and legends have become embedded in the larger civilisation, in which we now live.

So it's true to say that the very roots of Western Civilisation first began to grow in islands in the Mediterranean Sea about 2000 BCE, with the emergence of a Minoan society in the islands around Crete. There is not much evidence to guide us, but surviving fragments of the excavated palace of Knossos, reveal the ability to construct large gracious buildings, well decorated, and housing a well-planned water system. Have we discovered the earliest folk we could call European? Meanwhile to the north in mainland Greece, a second civilisation began to take shape some 400 years later. In this larger area, a number of individual city communities began to grow, exchanging trade and ideas with each other, and with societies further away in Egypt and Samaria.

Both of these small civilisations eventually declined though, and by 1000BCE, the area had lapsed into a Dark Age of which little is known.

However, 200 years later, the small city states began to flourish again, and a more lasting civilisation was born. Each town was built on a defendable site, with civic and religious buildings at their centres. The role of religion appeared to be to formalise and record the beliefs and knowledge of the communities, though worship of multiple Gods, in a system of testing for truth and factual credibility, developed by philosophers and historians.

Other branches of discovery and learning soon emerged, which later were to become the basis of more detailed knowledge in Western Civilisation. Among these were new mathematics, including algebra and geometry, astronomy, geography, history, medicine, and technology. Other arts and disciplines flourished including, drama, theatre, and the whole process of political debate and governance. This was overseen by often conflicting religious and philosophical debate, and this Greek world, with its literature, language and arts, reached the climax of its genius in the period 350-300BCE just at the time when a new civilisation was rising further to the west.

THE ROMAN EMPIRE

The Roman world was the second great system on which our own is based. Its strengths were military power, new communications, and meticulous planning, with a common language and government. It developed at the same time as the Greek civilisation, but outlasted it, and expanded to include not only Greece, but territories and peoples from the coast of Africa in the south, to England in the north, and from west to east, a 3,000 mile swathe of land from Spain to Syria. Under the rule of Augustus[1] and his successors about fifty million people were governed within one system—a factor which later was to accelerate the growth and spread of a new religion.

For 700 years its power and influence dominated this whole area, and to this day many of the ideas, institutions and insights of modern living has been based on the Greco-Roman world which preceded ours. It provided for later ages the first international language—with its roots still universally evident in science, law, and a number of major modern European languages including the Romance languages of Italian, Spanish, Portuguese, and in the Germanic group including German, English, French, and Scandinavian.

The Roman world also had a major influence in releasing the third great injection of ideas from the ancient world into Western civilisation. This was through its convoluted yet remarkable relationship with a small religious kingdom on the eastern fringe of its empire.

THE JEWS

In the 1ˢᵗ century BCE the Roman legions entered Judea, and an age of military dictatorship began with the appointment of a local ruler, Herod, in 37BCE as, 'King of the Jews'. The iron grip of Rome was to remain there tight and strong for 350 years. Upon their arrival in 168BCE, they came face to face with the culture of an ancient society, steeped in strong and unusual religious traditions. To the Hebrews, their tribal God Yahweh had been in a special relationship with them from the dawn of history. Their folklore, passed down through the generations, recalled that God had entered into a covenant with Abraham, the archetypal father of the Hebrew people. The agreement was that they would be his special people, committed to being his agents on earth, and that in return, he would guide, them, guard them, and lead them to stability, peace and happiness. The evolving folklore revealed that this was not an easy relationship. There was, for instance, to be much suffering for the whole tribe when they were trapped into slavery in Egypt, about 1500BCE, during the most powerful years of the empire of Thutmose II, the probable pharaoh in the time of Moses, one of the founding figures in Judaism.

Old Testament sources in the book of Exodus[2] describe repeated attempts to break free, with an eventual mass flight from the affluent Nile valley, down the north shores of the Red Sea, across the Sinai desert, and into the wilderness of what was to become their promised land—their home in Judea. Through the records in the Old Testament, we can see the adaption from life as nomadic warrior hunters, to that of a more settled regime as farming communities, growing crops and tending domestic animals. The literature of this evolving society reflected this in a number of new images of how their tribal God 'was'. He could be seen at work in providing food milk and honey. Metaphoric tales appeared of sheep, oxen, and at the other end of the animal kingdom, of struggles with wild predators, such as lions and tigers.

God helped them in their struggles with neighbouring tribes with clashes of language and cultural values. They used their faith in Yahweh to help them triumph over rival God figures, and the books of Joshua and 1 Kings echo the struggles with the regional rival God Baal, his armies, prophets and supporters. But Yahweh was still a tribal God, able to put an extra hour or so into a day, to enable the righteous to finish off killing the enemy, or to cause the walls of Jericho to collapse when the righteous shouted, amid scenes of genocide (supported by Yahweh) which are bizarre to modern minds, yet are statements of the harsh survival themes of the late Bronze age.

During a thousand years of life in the Promised Land, a vision slowly evolved of Yahweh being more universal— a God of Gods—and this was reflected in the growth of more specialised worship in the Temple at Jerusalem recording Yahweh as the creator and sustainer of the whole world. Some of the Old Testament psalms written in the 10th century, show this: 'Praise the Lord. Praise God in his sanctuary... let everything that breathes praise the Lord.'[3]

This widened awareness took root more deeply in the 7th century BCE, when the invasion from the north devastated Judea, killing most of the population, and sweeping their leaders and intelligentsia off across the desert, to far away Babylon—modern Iraq. There, for a century, the thinkers and leaders languished in hopelessness at their separation from Jerusalem—a symbolic disconnection from God, which they saw as a consequence of their own society's waywardness.

Yet even in this adversity, new growth in knowledge and thought began to appear. Much of the earlier oral traditions began to be written down, forming the basis of later Jewish and Christian faith. God, all powerful though he was according to Isaiah, a deity suffering because of the shortcomings of humanity. In other visions of prophecy, a humble suffering servant figure emerged to lead the people back towards God—

perhaps even physically back across the desert to their historic home. And in the mid–6th century this return occurred, and those who survived the long walk back across the desert to the west, eventually arrived in Judea, started to rebuild their homeland, and refresh the faith of those who had lapsed without leadership. More struggles occurred, which are reflected in the writings of Daniel, Zechariah and Malachi, and it was into this scene of feverish searching for national salvation, that the Roman armies poured, in the mid–2nd century.

They interrupted a process where increased awareness, and resilience, and practical wisdom had been transmitted down though thirty or more generations, and had produced a wily, wiry, vigorous society, able to survive the grimness and hardship of the surrounding ancient world.

A pattern followed of resistance to Roman rule, countered by savage repression, and the growth of hermitic sects in remote places, all seeking and planning final release from the yoke of the imperial intruders. This mixture of new salvation prophecy, the sects, rebellions and expectancy of the forthcoming end of time characterised the closing years of the BC era.

The First Century and the Rise of Christianity

The 1st century marked a turning point in the rise of western civilisation. That is now clear, seen from our own viewpoint two thousand years later. To understand its full significance, we have to zoom in not to Rome, nor to Athens, but to Jerusalem, on the fringe of great Roman world. There, the bubbling cauldron of Jewish frustrations boiled over, new prophecy emerged, and within two generations the message of Jesus Christ had taken root, and spread across the whole Jewish Diaspora.

Its savage suppression by the state and local authorities in many places failed to quench the enthusiasm of new followers, and eventually was unable to stop the rapid growth of the movement. By the start of the 2nd century there was a vibrant young Christian church, largely detaching itself from its Jewish origins, and beginning to spread beyond the bounds of the Roman empire— which itself was by then only five or six generations away from holding up the cross, and accepting Christianity as its state religion. Whilst through the course of history, most large movements had barely succeeded the death of their founders—Christianity was to become different.

Jesus was probably born in 4BCE, (during the local reign of Herod). Little is known of him until the last two years of his life, when he burst upon the scene. His charismatic preaching, healing, and opposition to the rigid temple based doctrines of the Jewish establishment attracted mass support among the poor and those living on the edge of society, and drew him to the attention of the authorities. Some of the writings of seventy years later flesh out the details, and include the strong oral tradition that itinerant movement leaders were at work in the troubled regions around Jerusalem, where Herod lived, and Caesarea, the Roman military centre and seaport. Some were based at Qumran,[4] a dissenting settlement on the shores of the Dead Sea, and others were itinerant, John the Baptist and Jesus among them. Some of Jesus' original followers–the apostles—appear to have switched from John's group to Jesus, who himself, by tradition, was baptised by John. The difference between the two messages is interesting. John preached the need to repent and prepare for the imminent arrival of God's new kingdom upon earth. Jesus, by contrast, extended this in two new directions. The promised kingdom was not a future one at the end of time—but was to be built *now*—in the hearts and through the actions of his followers. Secondly, it was to be accomplished through his own agency—a personal link with God, whose wishes for humanity it was his mission to accomplish.

The key points of his preaching and mission will be covered in Chapter Seven[5,] but for now the story moves on to the swift reaction of the authorities, with Herod obtaining Roman permission to arrest and try Jesus, the terror and flight of the disciples, and the rapid and violent execution just outside the walls of the ancient holy city. 'Truly this man was the son of a God' said the officiating Centurion, as he watched the death[6]—the start of centuries of meditation on the significance and purpose of what had happened.

From about CE33 to the end of the century, the Christian revolution unfolded—based upon the heartfelt feelings of his followers that Jesus had returned from death, appeared in person to certain of them, and had become available to all who approached him in faith and prayer. The followers saw him as being there to meet the challenges of their own lives and to prepare for his triumphant return at the end of the age, when God's final justice would be dispensed, and his intended new world established in the hearts of men and women. The actual historical evidence of this is of fluctuating value, and we are never likely to know the details. What we can see however is the amazing explosive growth of the new movement, and its separation from Judaism.

We can also see a change in the mind-set of the second and later generations when they realised that the triumphal second coming was not in fact going to happen in their lifetimes. Furthermore, the nature of the surrounding religious values was changing rapidly, with the elevation of the Roman Emperors to the status of Gods. So for the Christians the image of Jesus the humble prophet and movement leader (as seen by the earliest followers), soon expanded to present him as the cosmic Christ, present from the start of time, and available to men and women of all future ages as master and friend—as John later put it 'I am the way, and the truth, and the life. No one comes to the Father, except through me'.[7.]

Alongside this there grew a powerful belief in the nature of Christ's living presence in the hearts of his followers. The accounts of his resurrection and appearance to selected disciples spread throughout Christendom, and became the inspiration of the next decades. Soon, during the last thirty years of the 1st century, many of these were collected together by the writers of the gospels, and were passed down to following ages.

This transformation of the Christians can be seen in two ways—as something that occurred within them, or as a second revelation from God of his wider purpose, expanding the earlier images of the earthly Jesus. Both viewpoints are argued right into modern times, and need to be respected. The one can inform the other. The 'other' in this context is the view now growing among modern leading theologians that the original traditions collected together in the Christian gospels, were used by the growing Roman parts of the Christian world to reflect the need for a Christ larger than the Emperors, and became an embedded interpretation of the actual earthly dying Jesus, as physically rising from death— with the whole resurrection movement focussed on this occurrence centred in one body. This became the view of the western Church, and masked the actual view of the earlier generations that their resurrection was more fundamentally something which happened in the hearts of new followers, to transform them. Two prominent thinkers of the nineties, Dominic Crossan[8] and Marcus Borg,[9] having spent twenty years studying the textual evidence, then left their desks and moved to the sites of early Christian living, examining the architecture, archaeology, local museum evidence of the Eastern church tradition, and consulted its historians. This was vital work because of the rapid growth of modern civilisation wiping out the physical remains of the past with the hotels of the present. Its conclusions, now being written and published, strongly support the view that the energy of Christ was more truly to be found in transforming people and that this is available for today's church and world in a new understanding (for the west) of the available power of a creating, transforming God—to whom Christ is the signpost.

While this was growing, the more easily attestable historical facts of the all-encompassing Roman Empire, revealed the battle between the Jewish, and the occupying military regime. Revolt by groups of Jews grew apace in the sixties, and were savagely put down by the authorities, who carried out mass murders, and burned down a large area of Jerusalem, including the holiest of holies—the Temple.

This carnage had two results significant for the course of later history. The Christians split off from the Jews, and activists in both these movements fled from their homeland, to safer places of refuge. The separation of Christians and Jews occurred in the last years of the century for complex reasons—but among them was the pressure to assert the kingship of Christ over the claims of the Emperor, and of the multi deists of the Greek world. The escape of the religious of both faiths took them far across the empire, and to places beyond it.

In the case of the Christians, with the exception of a small group who remained in Jerusalem, the faithful scattered to Lebanon, Egypt, Asia Minor, parts of Greece and to Rome itself. The Jewish/Christian Diaspora grew larger. Soon the Jerusalem group were forced to move as well, with evidence in place names and folklore[10] that they came to the south of France, perhaps including Mary Magdalene and James the brother of Jesus.

During this period, with house churches growing in location and number, the Pauline letters were written (surviving evidence appears in Corinthians, Romans, Ephesians Colossians and other more individually focussed letters). The Gospels were written, and other non-biblical sources record the expansion and growth of the Christian movement. These include accounts by Josephus of the military action taken by the armies against Jews and Christians, and references by Tacitus and Suetonius, both Roman historians, to the strong action necessary in Rome to put down the 'followers of Chrestus'.

FROM THE SECOND CENTURY TO THE MIDDLE AGES

The Christians did not escape persecution nor hardship in their new locations, but the remarkable story of the following years, is one of growth and increasing stability. They flourished, just as the mighty Empire did, though the latter had reached the limits of its practical outreach. After a series of battles around its fringes a decline began, while the energy and vibrancy of the Christians began to appear as a possible source of strength to the rulers. Shortly before his death in 337, Emperor Constantine himself was baptised and Christianity became the state religion of Rome—a holy and unholy alliance of interests that marked the final decades of the Empire's existence.

During this time, in parallel, the Christians struggled to maintain a unified faith across huge distances and cultures, and many attempts were made to bring together a corpus of literature that would be accepted as God's word for both present and future.

During the course of the 1st century, some of the growing numbers of letters and gospel accounts were becoming recognised. By its closing years at least eight books were considered to be scripture and added to parts of the Jewish Old Testament to be used for study, teaching and worship. The core of a written New Testament was emerging.

By 363CE, the Council of Laodicea stated that only the Old Testament, along with the Apocrypha and the twenty-seven books of the New Testament, were to be read in the churches. Other Councils of Hippo and Carthage (393CE and 397CE) also recorded that the same twenty-seven books were authoritative.

There is much that could be written about the way Christianity grew in the next era of history, moulding the course of Western Civilisation. But for the proposes of our narrative, the general trend in the centuries which followed, was for the new religion to grow in numbers and influence for three or four more centuries, and then settle down into a new balance with the emerging societies. Unlike the eastern Christian world, the churches of the West were dominated by medieval Roman Catholic oversight, and the agents of the Papacy were in constant touch with the courts of local rulers. This relationship suited a world where the spirit of feudalism grew strongly.

Centuries later, in *Macbeth*, Shakespeare wrote, 'Come thick night... and pall thee in the dunnest smoke of hell. That my keen knife see not the wound it makes. Nor heaven peep through the blanket of the dark'. Historians of the early medieval years have echoed this by calling them the Dark Ages.

It was an era when the accepted political belief was that individuals were the chattels of local powers, who were approved by Kings, who were approved by the Pope, who was approved by God. Neat, but double edged, and a philosophy that suppressed universal enquiry, debate, learning and change. Life here was just a passage to greater things beyond. A period of static civilisation resulted, which lasted for eight centuries.[11] That does not imply that nothing happened in the centuries of feudalism. There were struggles, wars, famines, plagues and, around the edges of the new societies, there were continual problems with the different cultures of neighbours.

In the 13th century a series of military expeditions was mounted by western European kingdoms to recover the territories in Asia Minor overtaken by militant followers of the prophet Mohammed. The returning Knights of the Crusades brought back to their homelands plundered treasures and relics which became the prized possessions of rulers, and monasteries. These worthies also returned with practical knowledge from Asia Minor, of mathematics and civil engineering, filtered through from the Arabic world beyond. Many of the glorious cathedrals of Spain, France, Germany and England still bear testimony to this.

At the same time, Christian religious leaders began to refresh their faith with revised values, returning to the original message of Jesus, twelve centuries earlier. St. Francis of Assisi (1181–1226) became an icon for a growing number of followers—founding new religious orders, and planning ideas which centuries later were to be picked up by Catholic thinkers working to promote ecology—of which he became the patron Saint as recently as 1979. Fifty years after St. Francis, another great thinker, St. Thomas Aquinas, canonised in Avignon in 1323, refreshed earlier Greek philosophy of Aristotle with such clarity and logic that he too became a central theologian of Catholicism for the next seven centuries.

Pushed along by all these forces, inevitable change occurred and in the 14th century the medieval era finally came tumbling to its end. The doctrines of the church had become over-complicated with, for instance, the two sacraments of the early church growing to seven in number. Corruption and abuse of power had increased, and in parallel with these roots of unease, a new spirit of adventure encouraged explorers from Portugal, Spain, followed by France, the Netherlands and England, to sail away from the known world and find out what happened at its limits. Was the earth a flat dish? Would they drop off it? Although most intelligent thinkers no longer believed that, there was huge surprise at the returning evidence of a sparkling New World beyond.

Spanish explorers made their homeland the richest country in the world by returning with silver, gold, pearls and spices. Sir Walter Raleigh returned to Devon with vegetables called potatoes, and with an apparently pleasant drug called tobacco. Sir Francis Drake amazingly enough, circumnavigated the world in 1577, and returned with enough treasure handsomely to reduce the English national debt. The exploits of the intrepid explorers significantly changed trade and political horizons for the future.

The same spirit of enquiry flooded into the universities, and a reappraisal of learning followed—waters which eventually began to lap at the walls of the Vatican itself. A further and fundamental change followed from the invention in the 1440s of the printing press, making knowledge available to a dramatically wider audience than had ever been possible before. Books started to replace scrolls, and more people began to learn how to read, and then to write.

All of these came together to produce a flood of change— those who stood in its way would be drowned. The past could no longer withstand the present. The age of the Renaissance and church Reformation had begun.

————————

NOTES TO CHAPTER TWO

1. Augustus ruled as Emperor from 27 BCE till 14 CE. In common with other great rulers past and present, he was declared to be God by his followers, a source of huge friction between them and the Jews and Christians in later generations.

2. *Exodus.* This folklore is set down in *Chapters 3–10*, having been passed down for nearly a thousand years by oral tradition, before being written at length on scrolls dating from the exile of the Jews in Babylon (6th century BCE).

3. *Praise the Lord.* Psalm 150.

4. *Qumran.* To become famous in the 20th century when Bedouin shepherds discovered the first of the Dead Sea Scrolls in a cave just south of the settlement in 1946 (see *Bibliogrpahy*).

5. See *Chapter 7: Written Sources.*

References to the teachings of Jesus abound in the gospels, Pauline letters, and St Thomas's gospel, outside the NT canon. In particular note, Matthew 21–26, Luke 11–22; and for an unusual view, Thomas (114 claimed sayings of Jesus).

6. The crucifixion of Jesus is described in Mark 15:21-41. The Centurion's words are from Mark 15:39.

7. *The way, the truth and the Life.* See John 14:6.

8. John Dominic Crossan (b.1934). Former Irish monk. Biblical scholar, author and lecturer. Authority on historical Jesus and First Century Christianity.

9. Marcus J. Borg (1942–2015). Jesus and Biblical scholar, lecturer and author of 21 books. The focus of his work was often on what it meant to be a christian and that a deep understanding of the historical Jesus and the New Testament could lead to a more authentic life.

Borg and Crossan also collaborated on three books (see *Bibliography*).

10. *Diaspora.* However, many scholars dispute this, pointing out that the evidence of fleeing to France only starts to appear nine centuries later, and the earliest written reference is also from the same period in a papal Bull that was issued by Pope Benedict IX in 1040.

11. The Renaissance is generally considered to have started in Florence in the period 1350–1400.

CHAPTER THREE

FROM THE RENAISSANCE
TO MODERN TIMES

Stirring changes in western society began to burst out between the 15th and 19th centuries, beginning in the states of Italy, and soon echoed in many other regions of Europe. The word 'Renaissance' (French for re-birth) was applied much later, and described the flowering of Art, Music and other humanities which took place in these years, laying the foundations of modern times from the 18th century onwards.

CULTURAL CHANGES

At the heart of these changes was a rekindling of interest in the fruits of Greek and Roman culture from

1,800 years earlier. This was reflected in the works of Raphael, Botticelli and Leonardo da Vinci in the fields of painting, architecture and sculpture.[1]

Similar developments occurred in the field of music, where the original sole sources of patronage (Catholic Church and king's courts) became supplemented by secular concerts, theatres and bible based choral music in new protestant churches, all breaking away from the past. So the creative earlier forms of Palestrina, William Bird and Claudio Monteverdi were soon developed by Vivaldi and Corelli in Italy, the Bach family in Germany, and many others of the Baroque era including Purcell, Handel and Jean-Baptiste Loiellet, living in London. They all laid foundations for the later composers of the high classical era from the late 18th century onwards— including, at its start, Mozart, Beethoven and Schubert. Behind this growth of music making, were the skilled instrument makers, and the invention of new instruments need for their carrying power in the larger orchestra which started to replace baroque ensembles.

Changes in religious thought also gathered pace behind this, with strains appearing between the Pope and a growing number of reformers. In 1517 Martin Luther, a dissident Catholic priest, published his book *The Ninety-Five Theses*, which caused the beginning of a major rift within the church. Others including Huldrych

Zwingli and John Calvin joined in,[2] and in Germany new Lutheran and Calvinist churches were formed. Within twenty years, the church in England split away from papal control, under the marriage and succession demands of Henry VIII. With a split from Rome, the Church of England was re-formed in the troubled middle years of the 16th century, and soon after other groups appeared including Baptists, Independents, and Quakers. In later centuries the schism continued with the emergence of Congregationalists, Methodists and the Salvation Army movement.[3] While these changes swept through England, similar reformation pressures spread throughout Scotland.

But back in the 17th century, the beginnings of this were fundamental and often violent, with friction between the reformers and the traditional Catholics. Within this troubled period, many families in Southern England began to flee from the persecution, to start new lives across the ocean along the eastern seaboard of what is now the USA. Among these, in 1656, from a small village near the East Anglian seaport of Maldon were the relatives of a dissident local priest. Like many others they laid down new roots in Virginia, and three generations later, their descendant grandson, George Washington, became the first President of a new American nation. A long cry from the clamour in England at the start of this period when back in Essex, some twenty miles away from Maldon, a young man was caught quietly reading a Bible in the local chapel. In 1555, he had no right to be usurping the role of priests in this way, and was publicly burned at the stake. But within five years a school was founded in the memory of the youthful William Hunter. It still flourishes today, with the motto, *Incipe* (Begin). It was and is still true that people can be killed, but ideas cannot. The story of Western Civilisation.

ECONOMIC AND POLITICAL CHANGES

Much of this change was made possible by practical and technical developments in the background. The invention of printing largely attributed to Johannes Gutenberg in the 1430s enabled books to become widely available within Western Civilisation for the first time. In 1476 the printing press reached London, pioneered and financed by William Caxton. A continual flow of technical improvements made literature progressively cheaper. This inspired growth not only in the church, but more widely in universities, colleges and schools, in writing, and across a whole range of commerce and administration.

Alongside these were the progressive discoveries of 'new worlds' by explorers, between the late medieval period and 18th century. These started as early as the

13th century, when the Venetian, Marco Polo, following in his father's footsteps, journeyed eastwards, and reached China. Then generations of seafarers probed further, sailing to the edge of the flat dish that was the earth, to wonder if it was possible to look over the brim. In the late 1400s Magellan circumnavigated the world (though some still argue that this was not proven) and in 1497 Vasco da Gama found the lengthy sea route from the Mediterranean to India and the Far East. In later centuries, Sir Francis Drake and James Cooke were to extend the discoveries even further afield. All of these brought back wealth and treasure to European countries, and fuelled the growth of commerce.

Other implementers were the engineers who discovered new means of farming, and new ways to harness natural energy. Then there were the great contributions by capitalists, industrialists and landowners. Each one of the thinkers and the implementers quickened the pace of increasing knowledge, and broke away fragments of the older medieval images of God and Religion, which began to lose some of its credibility, and retreated to diminishing areas that it could still control.

The modern era was approaching.

Modern Times, Post Modernism and Beyond

From the 17th century to the world we live in today, a growing series of human achievements occurred—each episode creating another.

The starting point was a rapid change in the use of land. The old feudal system produced islands of cultivated land within a largely undeveloped (and lightly inhabited) countryside. New allocations of land made by a succession of Kings following the dissolution of monasteries, greatly increased the available farmed areas, with enclosed fields, steadily reducing the surrounds of forest, scrub and moorland. The basic modern rural road grid began to take shape, and can still be seen in patches today. Agriculture was also transformed by new tools and implements, and new forms of selective breeding of livestock over a 250-year period until the mid–19th century. It was a veritable Agricultural Revolution.

As I hinted in the last chapter, the thinkers and implementers were also at work in even more profound ways. The great expansion of trade, started off by the new world imports, encouraged manufacturing. Water power, and soon steam power, drove new machines in small factories for milling, making furnishing and clothing. The discovery of steam engines soon moved more manufacturing sites to places where coal, iron ore and copper could be mined. Whole tracts of rural England became suburban sprawl, and the population rapidly increased. (From four million in 1500, ten-fold four centuries later, to over forty million). This produced a similar order of social changes and pressures, many of them stoutly resisted by landowners and manufacturers. It was an Industrial Revolution.

The disciplines of Arts, Science and the academic institutions which housed them, continued to expand in this period—an era variably described as the Age of Reason, and the Age of Enlightenment. Great thinkers published their ideas—Newton, Voltaire and Rousseau among them. However, if the pace of change was rapid in these years, it was about to get still faster.

In the late 18th century, the transport of heavy materials had been improved by building a network of canals, ultimately linking industry to the highways of the sea. But in the early years of the next century, communications were transformed by the development of railways on which steam engines pulled passenger trains and freight. The opening of freight lines in Germany, and the passenger carrying Stockton and Darlington route in England (1825) precipitated a generation of railway mania. It was to change the face of Europe, and the lives of its peoples. In Britain, suddenly there was a standard time, a country-wide postal system and a national railway timetable. Newspapers from London could be delivered to remote towns every morning providing a nation-wide new service. Lobsters swimming around in the sea at Brixham could be eaten in Chelsea restaurants the same evening, and professional sport became possible with football teams from Lancashire travelling to play London teams on Saturday afternoons. Special trains began to carry thousands of supporters from one end of the country to another. The impact on society was immense, and by the time the railway network reached its peak in late Victorian times, it's worth remembering that many rural areas of England had better public transport than they enjoy today.

The communications revolution continued with the telegraph network, which quickly became Europe wide, and in 1858 extended across the Atlantic, reducing message transmission from ten days to minutes. By the late century, western societies had been transformed by these changes, and their underlying structures had been transformed—with the paternal oversight of feudalism becoming replaced by that of manufacturing capitalism.

The voices of mass populations were beginning to be heard, as witnessed by the extension of primary education to all children in Britain in the 1870s. Under these influences the ideas of 'Modernism', began to weaken, and new ideas began to reflect what society thought, and to drive it forward. Science had become the predominate source for truth and reality. The writings of Charles Darwin strengthened this enormously, and became a foundation stone for future science. Things that were measurable were considered to be aspects of the truth, but Religion and the morality based on it became increasingly threatened by the new thinking and relegated to the status of the subjective. Britain and other north European countries were becoming secular states.

But back to invention and discovery, to complete the story. In the first half of the 20th century, technology developed at an increasing pace in the west. New energy (gas, then electricity) replaced steam. Communications were transformed with the telephone replacing the telegram, and roads competing with railways. Radio and aeronautics grew out of the pioneers' ideas. In Europe the labour movement grew in the western regions, just as Marxism became imposed on the east. The role and employment of women was changed by desperate economic shortages. By 1917 they had the vote—and by the 1930s were beginning to enter the professions in greater numbers. The old order was collapsing.

As the 20th century unfolded, marred and crippled by the two World Wars, technical advances continued apace, with the further evolution of the motor car and aircraft in the first two decades, and radio and fast ocean travel between the wars. After the wars, huge changes occurred following the growth of mass television, and cheap air travel by a new generation of jet powered aircraft.

Atomic energy was harnessed to provide power, and to increase military strike power, and the birth of computer technology began to transform communications once again. Much of this was aided by discoveries in satellite technology, created in the wake of the American and Russian moon race in the 1960s. Not only would our grandparents have been astonished at the mechanics of the modern car, but also at its guiding voice—the sat nav—a product of the Space Race in the Cold War years. And as for mass package holidays in Benidorm or beyond and cheap flights over thousands of miles ... Well!

Post Modern Society

By the last quarter of the 20th century, the tumult of pressure was producing a marked shift in the values of society and the way it used new technology. The phrase 'Modern Times' used by historians was no longer adequate to describe the new scenery of society. 'Post Modernism' was the label picked up by historians, philosophers, artists, academics, sociologists and many others. Originally borrowed from the fields of architecture and design, the term came to be used more widely to describe the culture, values and chosen lifestyles of people living in the West today. At its heart it holds that there is no real external truth beyond what is created in the minds of people. It's the opposite of 'objectivity', which holds that the truth is external, present, always and something for people to discover.

It suited the spirit of the age, and spread to apply to all spheres of knowledge—including the sciences. Truth and reality were individually shaped by personal history, social class, gender, culture and religion. These factors, according to postmodern thinking, combined to shape the narratives and meanings of our lives.

But in some ways Post Modernism spread doubt and questioning compared with the human hope and optimism of the 19th century. It was secular. There was no single defining source for truth and reality beyond what individuals chose to believe.

This mood change had already been fuelled by the devastation and disappointments of two world wars. It then continued to grow in the post-war years, after the first euphoria of peace had faded. During the 1960s and 1970s, the prevailing attitudes against authority, institution and establishment all grew, and as this way of thinking progressed, it led people to think differently about popular culture, love, and marriage; and the switch in much of the Western world to the problems and opportunities of living in a service economy rather than a manufacturing one.

Beyond Post Modern Society: the Growth of a New Era

In the closing years of the last century, the realisation grew that the rate of technical change was so great that a new order was creating itself. Three decades occurred of ever increasing rate of change—exponential growth. New developments had emerged from the old, leaving the 1980s and 1990s far behind. The digital age was dawning, giving hugely increased instant access to knowledge. Computer power also increased exponentially, and advancing digital technology reduced the size and cost of computers, while increasing their capacity and speed. Whilst mainframe support for industry and administration advanced greatly in this period, the change was particularly marked at the personal end of the spectrum, with laptops outpacing PCs, and then themselves being themselves swept aside by tablets. Mobile phone technology blossomed, and began to threaten the dominance of landlines. At the same time, in developing countries further afield, the same digital revolution was already jumping far ahead of the older landline systems. This was bringing business and political opportunities to massive populations in Africa, the Indian sub-continent, and the Far East.

This emphasised a second consequence of these years of change, the growth of globalism. The speed and scale of communications began to present huge opportunities for worldwide commerce, making it possible for the produce on a British breakfast table to have arrived there from three or four continents. Not only did the digital transformation play a part in this, but also the growth of a worldwide network of jet powered air travel, carrying people and freight. This reduce journey times to a few hours, which only five generations before had taken months to complete.

Alongside this technical revolution came the human consequences. The rise of mass access to information produced a widespread phenomenon of pop power, based on individual assertiveness and increased interest in creative freedoms. Huge opportunities were suddenly available for those in a position to grasp them.

Conversely there were declines to be felt in trust, in authority, in local and national pride, in interest in conventional religion and many group activities.

Widespread local dramatics, singing in choirs, attending local public meetings or sporting occasions, are all examples of this change, while private and personal browsing round the electronic media have replaced them without people having to leave home. The long standing capitalist work ethic has faded in places, and large pockets of unemployment have caused painful social and individual problems for millions of people in the developed world, without distorting the huge differences already existing between their lot, and that of the emerging economies further away.

This all brings us face to face with our present—and a changing era for which there appears to be no label! That is not surprising, since every generation has found it hard to judge the significance of the 'unmeasurables' at the point where history and the current times join together. But a few observers of society have tried, including 'Altermodern' (Nicolas Bourriaud); 'Hypermodernity' (Gilles Lipovetsky); 'Automodernity' (Robert Samuels); and 'Digimodernism' (Alan Kirby). The writings of all of these brave observers and prophets are all worth further study. In the meantime, I will labour on and use the label 'Late Modernity'.

The issues we've considered here have now come to exert unstoppable pressures on politics and economics. Two other key factors now come into play; the growth of scientific knowledge mushrooms from astrophysics at one extreme, to micro, no, nano-technology at the other. Alongside this, conventional values and religious beliefs of post modernism are no longer tenable, turning society to other explanations and themes.

Taken together, these are the technical and material transformations of modern time. Such is the pace of change, there are many more, not least the astonishing progress of medical science in the last 150 years, increasing life expectation and the general physical health of millions of people in the west. But in parallel with these changes there have been many others in the less measurable heart-beats of society, in moral values, political beliefs, personal expectations, and concern for the effect of all this on our environment.

These are the key elements of taking stock of 'Where are we now?'—a topic to which we turn in the next chapter.

———————

NOTES TO CHAPTER THREE

1. Leading architects, painters and sculptors of the Renaissance include Leonardo Alberti (1404–72); Botticelli (in Florence) (1445–1510); Leonardo Da Vinci (1452–1519); Michelangelo (1475–1564) and Raphael (1483–1520).

2. Zwingli (1484–1531) and John Calvin (1509–64).

3. After an earlier start, Congregationalism began to spread in the mid–17th century; Methodism, following the Wesley brothers in the late–18th century, and the Salvation Army expanded quickly after 1865, in the footsteps of the dynamic William Booth.

CHAPTER FOUR

WHERE ARE WE NOW?

At the start of this book our story began with the violent windswept waves carving out a gash in a Suffolk shingle beach. The revelation of successive layers of old foundations gave us a glimpse of the past generations at work before our times, guiding us where to build ours. There was more to it than that, however, because the following day crowds of interested visitors arrived, encouraged by the TV reports. Was it a chance to see the old history of Dunwich revealed? No it wasn't, because a second tide had just swept along the coast, and all but covered up the remains once again. On the surface of things, there was nothing to see of the past, but underneath there was all the experience of past generations. Just across the wet landscape an old ruined farm building told the same story, with the tidal flood just stopping short of it, leaving a few cattle grazing safely around the ruins. Just up the coast more modern buildings were not so lucky. They had been built without an eye on past experience and they were now flooded. Six feet of disruption to suffer, wretchedness ahead, and bills to be settled.

There is a parallel for us. It is to realise that all those efforts of the past have put us where we are now. All the knowledge, the vast body of beliefs and values, all those ancient institutions, all the inventions, buildings, artefacts, they all constitute our world. We live surrounded by them. To become conscious of what they have given us is to begin to understand where we really are in our own life journeys.

THE STEPS TO THE PRESENT

The key to this new understanding lies in grasping how the thoughts and guiding beliefs of our western world have changed and taken new shapes for us all today.

To make the story easier to grasp, I see four steps on this journey. A sequence something like the following:

1. *Pre-Modern Era:*   1450–1850

2. *Modern Times:*   1850–1950

3. *Post Modern Times:*   1950–2000

4. *Late Modernity:*   2000–present

These eras and their titles cannot be precise—there are overlaps between the dates, and disagreements about the labels. But as we've made our brief journey through them, we may have seen markers laid down which will help us to understand where we are placed today.

Moving onwards from the late medieval period, for instance, we can trace the slow decline of the driving belief that the world was the centre of the universe, created and guided by an animate being—called God, for want of a better human term for something, someone so vast. For five centuries the predominance of this view drains slowly away, to be replaced by a more pragmatic, scientific rationale. We have also seen the results of this drift, forming the beliefs and values of a post-modern world from the second half of the 19th century onwards. Then, we have noticed the beginnings of an unusual reversal of this view, reflecting a desire to set people free to be creating discoverers of what is—the surrounding web of objects and themes *outside* our own consciousness. In parallel with this, within many of the sciences, belief is growing in the unity of a creative process through time, and of the presence of a vast yet meticulously detailed pattern present all around us—something greater than we have been able to see, sense or measure before. Lastly, the mystical religious thinking which recorded and controlled the earliest years of our civilisation, starts to reform itself into line with the sciences, suggesting a convergence in future.

This is a crude picture of the pitch upon which we now start to play. Its lines have been marked out by the past, but its goalposts have been placed by people in the present. The key question in this chapter is what sort of game are we having to play? What are its rules? Do we have to join in, if we don't like football?

The issues raised by this, fall into five groups. We'll look at them now.

What are the New Issues?

I. New Economic and Political Pressures

For many of us these are hard to study in a calm and rational way. It's one thing to talk about a global world, but another to realise what that means for us now, and in the future to come. It's not just the awareness of the sources of our breakfasts. Behind the consumption of the cereals, fruit and coffee on our tables there is a revolution of food distribution, involving supermarkets and suppliers in Europe, a network of international shipping and air freight organisations, and large numbers of wholesalers/ suppliers in far away countries. Add to that all the supporting industries, producing and servicing ships, aircraft and fleets of lorries and our picture of globalism starts to expand. Then add in the worldwide teams of experts increasing knowledge of soil propagation, of breeding techniques, of agricultural technology, and the complex picture expands still further.

This is not only a huge technological change, but also a societal one. Globalism involves massive co-operations between people from different regions and cultures, overcoming barriers of history, language and moral beliefs. The opportunities are immense, but the problems potentially disrupting and troubling for many groups of people facing changes in their lifestyles, employment and security.

In consequence, groups of workers in developed countries have to face unstoppable competition from labour in poorer nations, able to make their contribution for a fraction of the cost. The results are now well known, with re-training, long term unemployment, or moving homes, among the options. Economic uncertainty exists side by side with the growing wealth of those who benefit from new globalism.

This all manifests itself in new patterns of political activity, vying with older established political groups for the support of changed communities. And at an individual level, there is a breach of trust and belief in rulers and authorities, which we'll consider a little later.

II. THE POPULATION EXPLOSION

Growing alongside globalism there is another threat to the stability of societies, ours included. Awareness is spreading that the expansion of the world's population, will exceed the availability of humanity to feed it. Recent UN figures suggest a global population by 2050 of between eight and ten-and-a-half billion. In case 'billions' are concepts that are hard for you to grapple with—as they are for me—pause to look at one of the numbers another way; *10,500,000,000,000* people. Of these, most will live in developing nations, needing to be resourced by developed countries, whose own populations will be declining. As a result, fossil fuels, minerals, timber and water will be severely depleted in many regions.

To look at the UK figures is no source of relief either, with an estimated population at the start of our story, in what is now England, of about 6,000 people, rocketing upwards to 57,000,000 or more currently. In both the world and local examples, the present and future problems are produced by a growing life expectancy of the new-born and increased longevity of the older population. While the population projections increase, the actual number of jobs decreases through manufacturing decline and the transfer to cheaper labour markets in the global world.

The potential problems of these trends are compounded by a series of other factors, to which we now turn.

III. HARNESSING ENOUGH ENERGY

During what we have called 'Modern Times' the technical and lifestyle developments required vastly increased energy to be available, to power machines for new industries and for heating homes for a rising population. But enough was readily available locally—coal and iron ore for industry, timber for burning, and, later coal-gas for lighting and heating. Today that is not the case—local answers are not enough, needs have greatly increased, and with them the problems too. As we have seen, advance predictions for the future supply of oil, coal and minerals do not match the anticipated demands of a developing world, and an increasing population. Further attempts to harness the power of wind, water and sunlight have made progress but not quickly enough. Nuclear energy has a long list of potential risks attached to it for the future. So there is a growing potential crisis for us to consider now, and for our children to address in the future.

In terms of statistics, world needs for liquid energy will, by 2040, have sharply increase from 170 quadrillion Btu to 230, coal and natural gas from 260 to 410, renewables from 55 to 120, and Nuclear energy needs from 25 to 55. Note that the required present energy, and more so in the future, is for fossil fuels, which will still exceed most of the alternatives added together. Consequently, the depleting supplies of coal, oil and gas will increase in price, and place more power in the hands of state or commercial providers. This increase in energy demand will also add to the next problems to mention, those of climate change and atmospheric pollution. The remarkable agreement on resources between 185 countries, shows a greater awareness of an impending world crisis, but the will-power to avert it has now to be tested in the future.

IV. CLIMATE CHANGE

As this is an intensely politicised area, the following definitions might be helpful.

- *Weather:* The condition of the earth's atmosphere at a particular time and location.
- *Climate:* The condition of the earth's atmosphere over a long period in a locality or wider region.
- *Climate Change:* A long-term change in the Earth's climate or of a region on Earth.
- *Global Warming:* The increase in Earth's average surface temperature due to rising levels of greenhouse gases.

Although little is known of the Earth's climate over the spread of recorded history in the last 5,000 years, it is possible to form an image of fluctuating conditions that affect where people could live, and find or grow food. Further back, in the last 650,000 years, geology reveals to us seven stages of glacial advance and retreat in the middle latitudes, making life barely possible, with recoveries and more life evolving in the gaps. And of course before that in the preceding eight million years there was a constant flow of climate change as the molten blazing Earth slowly cooled down.

For us today, a key question is whether the present signs of climate change are just normal and small variations in a long term process, or an unusually rapid onset of developments caused by natural events or human actions. Recent evidence revealed by NASA claims that, 'Ninety-seven percent of climate scientists agree that climate-warming trends over the past century are very likely due to human activities', and most of the leading scientific organizations worldwide have issued public statements endorsing this position.[1]

The increase of Carbon Dioxide into the atmosphere in the last 150 years has trapped heat and raised global sea levels in excess of 17cm, while global temperatures have significantly risen too. These have warmed land masses and the top 700 metres of oceans. The combined effect of these factors has decreased the size of the two polar ice sheets, with Greenland losing up to 60 cubic miles of ice between 2002 and 2006.

Lastly, and significantly, there has been a measurable increase in extreme weather incidents worldwide, in the last thirty years.

V. Atmospheric Pollution

This is being caused by spreading chemicals or biological matter into the earth's surrounding atmosphere that are harmful to living creatures. But is this a natural sequence of events, or one caused by human activity? Volcanoes or power stations?

Since the beginning of the Industrial Revolution, the acidity of surface ocean waters has increased by about 30% due to the emission of more carbon dioxide into the atmosphere. The amount of carbon dioxide absorbed by the upper layer of the oceans is increasing by about two billion tons per year.[1] The Earth's atmosphere is precious, special and life giving. The evidence seems clear that the humans are destroying it.

Who is Responsible?

Who takes responsibility for each and all of these issues? What is more, who *needs* to take responsibility?

Does that just involve governments or others who we will never meet, or does it involve *us*?

A key word here is 'Stewardship'. Sensitive care for the world's resources, for human invention and intervention, and for the values and practices within societies. This leads us to a completely new area within the mantle of stewardship, the role of human beings as creators. This vastly increases the opportunities and knowledge available in the age of Late Modernity. So who takes responsibility for this? Who will pick up the mantle of passing on our changed and extended knowledge to those who will follow us?

All of this places us in an uneasy position, as we contemplate our future approaches to the steps we take in future. But then the analogy of having to play football was not easy in the first place. The next chapter will ask further questions about playing this game. Its title is, 'Where are we going?' and its questions are: What are the skills we'll need get through the challenges? What are the strategies we will need to employ? What do we believe it's all about? And if it's about winning, what are we trying to win? What are the societal responsibilities, and what are ours? Will it help our own journeys to pick them up and personalise them?

———————

NOTE TO CHAPTER FOUR

1. *Climate Change*. See climate.nasa.gov 2015

CHAPTER FIVE
WHERE ARE WE ALL GOING?

The opening words of this book invited us to think about our own story, the journey we're taking through our own lives. It asked, 'Are we just swept up in it, like pebbles being washed along on a beach by the power of huge waves? Do all the circumstances around us just sweep us along, without us being able to do much about them? Are we in fact trapped by these huge pressures and unable to escape from them?' Do we accept the flow of the tide, and just take from it such comforts as we can find when they appear along the way? Perhaps the comforts of money, security, help and protection from others? Or are there other ways in which we can free ourselves from the waves, and start to rebuild our lives, giving us more control, growing a new sense of purpose, and finding more depth and enjoyment in what we're doing?'

History suggests that there are indeed other ways, whether one bases trust on religious or secular values. Two things we have by now discovered about ourselves from earlier chapters can help us to re-focus our views. One is to apply the experiences of past generations to the decisions we now have to make in our own times, and the other is to think through what convergence of recent scientific and religious thinking may really start to mean. Both can profoundly alter what we put in to our personal future planning, and what we get out of it.

EXAMINING OUR ATTITUDES

History suggests that while the context of each generation alters, the range of human reactions in each era is remarkably similar. We have fears about survival and security just as our predecessors did. The instincts of sex, hunger, and group behaviour are in us all, just as they were within Stone Age men and women. The desire for order in society, with just rewards and opportunities, are not emotions we have suddenly discovered. Neither is it a new instinct to search for whatever is beyond the current horizons of human knowledge. It's one that's driven people forward since the dawn of recorded history.

So the basic processes of receiving from the past, and reacting to its implications are worth a further look. In Chapter One we looked at the role of each generation in receiving a bundle of inherited values, adapting them, adding new information to them and passing them on to the next era. How can we use these processes to guide what we choose to do today?

A starting point is to analyse why a particular issue or theme really bothers us. A first question is, 'What is the main background to our problem?' In recent years for instance, much public disquiet has been caused by evidence appearing many years later of sexual exploitation of young or sick people by predators whose activities were not 'noticed' decades ago when the offences occurred. How could this have happened? Is it important to discover the background, or personally best to let sleeping dogs lie? Viewed by the standards of today's society it soon becomes clear that one reason why the past must be properly scrutinised by the present, is to prevent similar activities now continuing unchecked because our own levels of critical awareness and sensitivity have not increased enough.

When the historic misdeeds are analysed, there are motives of offenders to consider, and also the alertness of society around them. The human motives involved can often be the same, generation by generation, whether psychiatric, childhood conditioned, calculated lust, compensation for insecurity, or inherited societal attitudes. Observance of how this has affected people in the past can help us to understand what is happening in the present. Similarly, people in decision taking situations must ask questions of their own systems and practices, lest weaknesses in them remain out of sight and unconsidered.

It may all have happened before, but everyone has, as an individual, a role to play in preventing its occurrence in present and future. We have to develop the sense that the past surrounds us with its vocabulary and values, and our starting point for modern life's journeys inevitably spring from them.

This will inevitably cause us to question and reject parts of our inheritance, and adapt others. Today's world is no longer inclined to accept lifestyles based on male domination, racial, religious or class assumptions which were acceptable to previous generations. But changing these does not happen easily or universally just because the tide is starting to flow in a new direction. Once again, individuals have to make up their own minds about what they need work for, and what their new beliefs, hopes and ambitions really are. The more honest we can be about these, the quicker the changes will come about.

Two things we have by now discovered about ourselves from earlier chapters can help us to focus on the way ahead. One is to apply the experiences of past generations to our present future options, and the other is to probe a little deeper into the convergence of recent scientific and religious thinking.

A final attitude to shape for ourselves is to decide what we wish to pass on to those in our families and other people who will succeed us in future. Chapter One recalled that no civilisation can continue to grow without the main body of acquired wisdom being passed on. History shows that positive present steps have to be taken to make this happen. It's not enough to make records, write or create art—action has to be taken to pass them on. It described details of three ways in which this has been done in the past through passing down specialised knowledge, widening its availability in the present, and taking care in handing on family or tribal values. Every age wants its children to be shown what the past has learned about healthy and good living. Not to do this is to start a downward plunge in social cohesion, and when the young have lost their way in the past, huge problems have followed.

But now we turn to the second theme we have discovered from the past. It is to consider something that thinkers are now starting to examine in more detail, an apparent convergence of recent scientific and religious (or mystic) approaches to human knowledge

The Convergence of Recent Scientific and Religious Thinking

The attitudes, beliefs and practices of scientific thinking have increasingly dominated western society for seven centuries. As we have seen in earlier chapters, this has progressively widened the gap with the religiously based beliefs of earlier ages, from which science itself originally grew and detached itself. The gap between these two produced times of struggle and tension, before lapsing into the shadows as our secular society more firmly established itself.

But these still remain the two all-embracing areas of human knowledge. What is more, they still continue to exist side by side, and each continues to research and develop its own frontiers—though independently and largely without consideration of changes in the thinking of the 'other side'. However, in our own attempts to be sensitive to everything that is relevant to how we face the future, one new factor is there to be considered. There is an emerging scientific view that shows openness to the existence of something 'out there'—an external creative force—not just a construct of our own minds. Similarly, in the other camp, there is a growing interest among liberal religious thinkers in the recent work of scientists, and an interest in studying how these new models influence their own ones. Let's examine some examples of this process at work—science first.

COSMOLOGY

To start with, consider cosmology, the scientific study of the large scale properties of the universe. Our knowledge of the universe around us increases at a staggering rate. Although much research is now more openly publicised—a feature of the new world of Late Modernity—it is hard for the non-specialist to keep up.

Within the last twenty years the view has emerged that 13.7 billion years ago a creative process took place—and is now adopted into public currency as the Big Bang. To get an image of the vast time and distance, it's worth remembering that a billion is a thousand million. Then ten billion years later the first stars and galaxies emerged, followed by our Sun being created four billion years ago, and the first elements of the solar system half a billion years later, including the planet Earth. Our evidence on the main shape of this process is based on the fact that we can see a number of key events now, due to the long time-delay of information reaching earthly observers.

Zooming in on the timescale a little more, plant life emerged, and much later animal life appeared, with mammals being traceable from sixty-five million years ago. At the seven million years' mark, bi-pedal creatures, and later, hominids appeared, with ancient homo sapiens appearing 500,000 years ago. Summarising the whole of this vast span; recorded human activity began in the last few seconds of this process, about 5,000 years ago, giving us the baseline from which historians can start searching for the roots of all our current knowledge.

In trying to understand this process, two developments in scientific understanding provide some explanations. The first is based on Einstein's theory of gravity, originating from 1916, showing that while Newton's original 17th century description was satisfactory for still or slow moving objects, it needed to be expanded for a wider and faster context by seeing it as a distortion of space and time itself.

A second explanation applies these dynamics to the universe as a whole. As the NASA website puts it, if one viewed the contents of the universe with sufficiently poor vision, it would appear roughly the same everywhere and in every direction. That is, the matter in the universe is homogeneous and isotropic[1] when averaged over very large scales. This is called the Cosmological Principle. It is an assumption is being tested continuously as we actually observe the distribution of galaxies on ever larger scales. In addition, the remnant heat from the Big Bang has a temperature which is highly uniform over the entire sky. This fact strongly supports the notion that the gas which emitted this radiation long ago was very uniformly distributed.

From these two themes a number of theories have developed from the very large explosions which have created the heavy elements of the universe—iron for example.[2] A recent book by the Italian physicist Carlo Rovelli,[3] aimed at an interested non specialist audience traces how physics in the 20th century contributed to our world view, and continues to develop and shape our thoughts today.

BIOLOGY

In perceptive writings on new ideas in biology during the last two decades, Cambridge academic Rupert Sheldrake[4] argues that present physical life forms, whether animal or plant, are supported by specific information fields created by other life-forms in the past, which they draw upon to defend, guide and support themselves to cope with the present, and to reproduce themselves, and evolve further in future. He describes this as morphic resonance, where past forms and behaviour of organisms influence present ones through direct communications across time and space.

As long ago as the 1920s experiments were beginning in this area. In Southampton University UK, for instance, the behaviour of Blue Tits was observed, tearing the tops of milk bottles on doorsteps and drinking the cream. Soon it became apparent that this skill showed up in Blue Tits over a hundred miles further away, even though their normal local range is under fifteen miles. An expansion of this practice was then observed by ornithologists wider afield, until in the 1940s it was seen to be universal throughout Britain, Holland, Sweden and Denmark. Interestingly enough, in Holland, there was a lapse in the data for eight years when the German occupation cut off milk deliveries (five years longer than the life of a Blue Tit.). However, when in 1948 the milk started to be delivered again, within months these little birds all over Holland were drinking cream, a habit that had taken decades to take hold before the war. How did they get this knowledge?

For that matter, how do birds newly born in the UK find the instinct to feed themselves up, and then to set off on huge journeys to central Africa, to arrive at a specific group of bushes that their parents came from the year before?

What are the human parallels to this? American writer Judy Cannato echoes the idea that information fields exist to support humans in this manner, and draws parallels to the writings of past thinkers two thousand years ago, describing a similar sense of external support, using the simpler language and imagery of their own times. Can the past be available to support the present in this way? she asks.

In similar vein, medical journalist Lynne McTaggart draws together the experiences of other scientists in developing the idea of a sea of energy, a cobweb of energy exchange, reconciling mind with matter, classical science with quantum physics, and science with religion.

MEDICINE

Quite apart from Cosmology Physics and Biology, other examples of this new thinking exist across the whole scientific spectrum. In many major areas of human knowledge, the boundaries have progressively been pushed back, alongside evidence of an interface appearing between the advancing researches of biology, medicine, and engineering on the one hand, and religion and philosophy on the other. (Of course this interface spreads even wider still into a whole range of arts and humanities).

As an example of the value of opening our minds to the widest range of possibilities it's interesting to look at some of the boundaries in medicine and see how the experts are exploring their edges. For instance, what can be gained by examining medical practices and beliefs wider than those which have so successfully advanced diagnosis and treatment in the Western world? What of the models and processes of oriental medicine? Many of these find analogies for complex cosmic events in the simple examples in life and natural; events all around us now. All the different things in our world are really just parts of the same whole.

Parts of this approach extend to Acupuncture, homeopathy and herbal medicine all of which are practised in Western civilisation, and may be considered under four application modes; the Physical Body, the Body and Mind at work, Body Energy, and External Energy. Thus in homoeopathy examples of treatments can be drawn across these four areas with the material content of remedies progressively diminishing, and their potency increasing. A similar pattern can be discerned in western medicine, using the analogy of the same four areas of treatment.

Changes in Scientific and Religious Approaches to Human Knowledge

However, if we switch the focus to the world of religion, we see the same thematic development at work. Medicine may show a progressing approach through Surgery,

Psychiatry, Radiology and external micro energy, but the theologians and philosophers can point to their new understanding of historical practices (liturgy), mind and memory, the energies of prayer and contemplation, and the presence of external power and energy in human experience. The models, images and language may be different, but the human knowledge in all these areas is converging, laying out a fundamental background to our planning for the future.

So as these different approaches to our life issues start to converge, where does this situation leave us as we start to consider our own ways forward, both as individuals and as part of a great surrounding society? What are the options—the alternative ways for us to make a difference to our worlds? The next chapter asks what part we can all play in contributing to the future we would like to see. We'll look at these through the lenses we've now identified of secular thought, and religious beliefs.

NOTES TO CHAPTER FIVE

1. see *Glossary* for definitions of Isotropic, Homogenous and Homogeneous.

2. see *Bibliography* for further reading and browsing suggestions.

3. Carlo Rovelli, *Seven Brief Lessons on Physics* (Allen Lane, 2015).

4. *Morphic Resonance.* Rupert Sheldrake first put forward his hypothesis in *A New Science of Life* (1981) and discussed it at length in *The Presence of the Past: Morphic Resonance and the habits of Nature* (1988).

Morphic Resonance: The Nature of Formative Causation (2009) is a revised and expanded edition of *A New Science of Life* (see *Bibliography*).

see www.sheldrake.org/research/morphic-resonance

CHAPTER SIX
THE FUTURE—
FLOURISH OR DECLINE?

Having considered the great sweep of history that has given us all our values, and placed us where we are, how do we use this information to guide us into the future? Having realised the huge pace and depth of recent change, how do we swim within this tide, and find the security, happiness and fulfilment that we all want in our different ways? What about the future—for our world, for our great society, for ourselves as individuals with new power and opportunities within it?

Secular Society

Our starting point must be that Western European governments follow secular agendas, which are supported by most of their citizens, who base their values upon them. So, from that perspective, what is there for us to do? The first point to make is that many repetitive episodes suggest that when societal crisis seems to peak, new ideas emerge. Examples abound. When the first and weakest Minoan states began to decline after 1000BCE, society was rekindled by local efforts with the growth of city states, each encouraging trade, and the growth of knowledge. This then linked together to produce a broader, stronger society—the Greek world. Although this growth was gradual, over a number of generations, the new was eventually much greater than the old.

Then, moving to the late middle ages, there was another similar episode. When the energies of the medieval era began to weaken and stultify, new ideas began to emerge from discarding some of the old ones. So new thinkers inspired a revived spirit of enquiry and adventure, with explorers and scientists crossing new horizons, escaping from the cloying culture of Church and kings. The pace of change had increased, and this transformation took less time than the Minoan one.

More recently, there was another example among many—in the last two decades of the 20th century, parts of Post Modernism were broken through by the needs of emerging sciences to free themselves from older, more restrictive dicta. The information explosion and growing feeling of personal independence grew stronger in the wake of this, and this latest process of change is taking even less time, indicating once more how the pace of life is quickening as the accumulated knowledge within society grows.

From all of these quick glances back over our shoulders, we see that certain characteristics of change have been common to successive ages. We start to sense how the whole long-term values of human life through the centuries have remained remarkably similar, although present circumstances seem entirely new to us.

People who've gone before us have found the courage to modify their inherited values, and at other times to reject parts of them. This required awareness of what had changed, and positive suggestions of new ideals to be grafted onto the past. Thus in recent times, older religious ideas of charity towards neighbours were replaced by human rights principles, emerging from the end of the two World Wars. They demonstrated the justice, compassion and freedoms that we all sought. These principles were credible to individuals, as well as larger organisations or governments. Ability to engineer change is essential to growing societies and individuals alike.

Another lesson from the past, is the element of risk-taking to create change. Risk taking was more normal for past ages than it is for us. Marco Polo walked two thousand miles to China, to develop what his father had started. Martin Luther dared to confront his God without going through intermediaries within the established Church. Rousseau risked rejection by his peers and death for expressing doubts about the old social order, just as the French revolution boiled over. All of these created enough freedom around them for fellow questioners to re-appraise what was true in their own heritage, and what needed to be expanded to widen and re-shape current beliefs. Many of these reformers found comfort in groups, and many of them were leaders of great intellect and demonstrable courage.

But what is new within our present generation, is the hugely increased opportunity for ordinary individuals to institute changes themselves—perhaps locally— perhaps on the wider stage—and even globally. Many examples of this occur all around us routinely, and each one encourages change, and has the potential within it for increasing hope, and for improving our society. Much of this new power to individuals comes from the digital revolution. The petition, the blog, the sudden interest of the media in 'human stories', these are all powerful pressures which individuals can harness from home without specialist help, or the need for detailed technical or academic knowledge. These initiatives frequently show the power to trim and change the published policies of leaders. This phenomenon has created a new ferment of change, inconceivable ten years ago when a global popular internet was beyond imagination.

A completely different challenge, new and never known before, is the ability of humans to move beyond this and become creators themselves. Whilst all previous generations have been able to modify natural resources, and forms of life—we uniquely have found the ability to engineer new life forms. As individuals we may not have the specialised knowledge to do this ourselves, yet we are able to have a voice in how they are used. Can we—used in the widest sense—create new food to feed the overpopulated world? Can we modify cells produce new ones to change the shape of human inheritance, perhaps to eliminate anti-social behaviour? Can we create robotic creatures to do routine work for us? Can these create their own decisions and carry them out? Someone has to decide, and today's world begins to require that there is a place in this for you.

Our own personal challenges have therefore changed in this latest, fastest development in our context. As we look again at the six challenges for us, which emerged in Chapter Four, can we redefine our approach to each one as an individual? As we reconsider the second and third issues can we form our own views on how our neighbourhood, country, or global world should act? Lastly, for we live in the age of the sound bite, can we succinctly, in a sentence express our position on each issue—writing it down so that when opportunity arises, we can remember what we have agreed with ourselves! Our resolution and clarity is needed to help others, and to maximise our own satisfaction and self-confidence.

A ROLE FOR RELIGION?

As a society holding pragmatic, secular views, when we come to consider our future challenges and opportunities, is there anything else that might reveal different dimensions of truth? Is there a role for the views of the religious? While they are now a minority who share many of the secular values themselves, are they seeing an extra aspect of reality beyond their own reach and thinking? This is an approach which may be held by people of religious backgrounds or none, and it would be tempting to list here the contributions that could come from atheist or agnostic mystics, as well as the whole span of religious opinions in our western society.

But such a review is outside the span of a book like this, beyond mentioning with respect where one can look for these views. In particular, there is much information available on Judaism and Islam from the Middle East, and Buddhism, Hinduism from further east. All these major religions acknowledge the acceptance of an external creating, active and empowering force, to whom they can relate and link in a range of different ways. A little research is rewarding as one realises how much they share in common.

But within the context of Western civilisation, so much of our background is based on the Christian heritage, that it has to be the point from which the general religious input is best illustrated.

———————————

CHAPTER SEVEN

THE INFLUENCE OF CHRISTIANITY ON WESTERN CIVILISATION

What extra dimensions can this particular heritage add to the mix already established as the major current secular view in our society? How does such a different view extend the majority view? What new flavours might it stir into the mix of our future decision making?

I suggest there are a couple of major distinctive departures from secularism; a belief in an external guiding presence in the universe, and a desire to refer to a specific store of time-tested human wisdom for guidance.

An External Source of Truth?

The belief that there is an external truth, separate from our own imaginations, has been given impetus by the changes in thinking over the last two decades, and the convergence of scientific and religious approaches. Its adherents ponder the presence of an external force, far removed from the earlier historic images of a tribal God, or a single living super-being, with controlling reins attached to human destinies. Yet for all its greater modern sophistication this view shares with earlier ones the problem of finding the words to describe it. It is of course impossible to use human language for something so far beyond our ability to measure or describe. Although the last chapter began to consider this from the secular point of view, the present one will return to an age old word; 'God'. I appreciate that this concept will make some readers uneasy, but think it's worthwhile using, to fully understand how the Christian view has built up, tested and developed its meaning through the centuries.

It is a view which sees God as a guiding, creating presence wholly outside the constructs of our own bodies and minds. A benign presence that is alive and the source of all matter and life within the universes. A power that is caring, evolving, changing. Words cannot fully express it, but the view accepts that God can place information within our minds, individually or collectively, and that something within us enables us to respond.

These human responses take the form of acknowledging the presence, the vastness of the whole created order, giving thanks for the special exuberances of life and of all we have and hold. They also draw out from us our frailties, worries, doubts and needs, and our requests for forgiveness for our shortcomings. They lead to further requests for insight and strength to deal with challenges before us. They draw out from us requests for help from God's greatness for other people whose problems we feel and want to share. Google the Christian meaning of 'grace'. This whole vast subject of the nature of God has its own specialised sub-themes and its own technical terms. I pass them by at present, to keep the main theme flowing.

A common feature of many major religions has been the need to personalise our images of God, and this has often been done through the imagery of God as a human figure, or as seeing the powers and qualities of God through the life and actions of a great historic figure. We've seen already how a range of Greco-Roman deities illustrated this phenomena, as did a number of examples throughout history, and which are around us at present in today's major faiths.

So for Christians, the example of Jesus, living over 2,000 years ago, set before his followers a model of qualities which came from God beyond. Through his own example he called them to lives of giving and serving, which could set them free from the traps of worldly situations to find the greater freedoms of closeness to God. It offered them transformation—a state of bridging the gap between forms of life here on earth, and beyond. Guided by these energies and insights the Christian movement grew through sixty generations of change and have moulded the shape of Western civilisation.

This was based on the existence of two unique factors, a body of special writings, and the growth of doctrine in the many churches which developed and evolved throughout the following ages. To understand these main themes is to gain further clues to the questions of where we have come from, and where we can go. We'll look at them in turn.

WRITTEN SOURCES

Christian Scripture has its roots in the early writings of the Jews, and comes to a climax in the events of the first three centuries of the common era—a span of about 1,700 years. Within the thirty-nine books of the Old and the twenty-seven books of the New[1], there is a treasury of guidance for good living, how to sense God, and the great beyond which human minds find so hard to probe and understand. They cover various concepts of God, ranging from a Creator, a Father figure, to a living spirit all around us and within us. From its many allegories parables and illustrations, different ages have drawn the wisdom that best fitted their own needs, and succeeding generations have remoulded these to match their own fears and hopes. So as we have seen, rejection and re-selection has become a prime means of finding the guidance that best suits us.

From all the themes that have been pondered upon through the ages, I choose five which speak powerfully to the 21st century.

I. SHOWING LOVE IN YOUR LIFE

The Two Great Commandments: These appear in Mark's gospel, with later copies in Matthew and Luke.[2] Written over thirty years after the earthly Jesus lived, they appear to record an anecdote passed down orally by the first generation of the followers who knew him. The theme is that an educated critic of Jesus, resenting his popular appeal to the common people, tried to trap him into quoting scripture which could incriminate him—perhaps capturing him breaking the strict religious conventions of the day—by asking him 'Which commandment is first of all?'

But Jesus side-stepped the trap, and responded instead with two different commandments from the past.

1. Deuteronomy 6:4-5—Love The Lord

'Hear O Israel: The Lord our God is one; you shall love the Lord our God with all your heart, and with all your soul and with all your mind, and with all your strength.'

Later ages have pondered this choice and expanded its meaning. The call is to focus on one creator, in human terms perhaps a father figure. More than that, it is to establish a relationship, a two-way flow, as implied by the use of the word which we have to translate imperfectly as 'love'. This focus had to be at the heart of the follower's personality, acknowledging the great power and care of the creator, and the tiny significance of the follower—yet able to sense God and respond to a living presence within him/her, or surrounding them.

2. Leviticus 19:18—Love Thy Neighbour
But while the first commandment may have been good enough for the priestly practitioners, more was required by the second selection made by Jesus.

'You shall love your neighbour as yourself; there is no commandment greater than these.'

The Christians had to be within society and as their relationship with God led them to live their lives a certain way, so this had to flow through all their dealing with people around them. These were 'neighbours' in other words not people they had chosen, but people who were just there, next to them. Other key examples of Christian practice take this further, because the ancient Jewish scriptures had sensed a priority that God appeared to favour people whose lot was harder than it was for those who were nearer the top of the pile.

An interesting corroboration of this theme comes from a contemporary anecdote about Hillel, a rabbi at the same time that Jesus preached. He too was asked a trick question about what was the whole of the Torah, the collected Law of the Jews. He replied:

> 'What you find hateful do not do to another. This is the whole of the Law. Now go and learn that!'

II. DAY BY DAY PRAYER

The Lord's Prayer:[3] Underlining the emphasis on giving of oneself to others, is another central thrust of the Christian message, a few snatches of evidence in the gospels, showing Jesus' perspective on how God calls followers to live out their lives. The actual words were only written down thirty years afterwards, with the earliest of them, in Mark's gospel simply proclaiming:

'Whenever you stand praying, forgive, if you have anything against anyone; so that your father in heaven may also forgive you your trespass.'

Luke, however, expands this to make five points:

1. Father, hallowed be your name.
2. Your kingdom come.
3. Give us each day our daily bread.
4. Forgive us our sins, for we ourselves forgive anyone indebted to us.
5. Do not bring us to the time of trial.

So the most familar version (which has nine points) does not appear in the earliest documents—we have to wait for Matthew's writings for this. Let us stay with the early sources.

If we think through these two versions carefully, a picture emerges that helps us build a constructive outlook for society in the immediate future.

In Mark's gospel the reference to 'standing and praying' is an exasperation oft repeated in the early writings. It reflected that the established religious Jewish people liked to appear in public as respectable role models of how the rest of them should live. But Mark's interpretation speaks to the present, and in effect says, if you want to adopt a public position, start in humility by exposing your own prejudices, and forgive those who have obstructed, frustrated and hurt you in any way. A good way for leaders, teachers and politicians to remake the new starts they claim today. A good way for us all to follow in the minute detail of our own relationships too. Which implies that *only* if one breaks through that barrier will a close relationship to God become possible. As the 1st century writer puts it archaically, '... so that your father in heaven may also forgive you *your* trespass'.

Let's look at Luke's expanded version point by point:

1. 'Hallowed be your name.' Hallowed is translated from the same word that in other places is translated as 'sanctified.' So, we begin the prayer by praising God. (This phrases also implies the commandment not to take the Lord's name in vain.)

2. The mysterious phrase, 'Your kingdom come' is so full of deeper meaning that it needs an entire section of its own to be able to explore it. See IV. Wisdom from the First Century, later in this chapter, and you'll discover some of the significance behind this simple phrase.

3. 'Give us this day our daily bread' is a theme which the 1st century readers would have understood as a repeat of earlier history, where with God's guidance, Moses, in the 14th century BCE had led the Jewish tribe escaping from Egyptian slavery. They edged towards a promised land (later, Palestine), finding that the journey was desperately hard, and starvation threatened the moving population. The pages of Exodus record ways in which the tribal God helped the suffering people to get through, an experience which lasted many generations.

For our generation there are two messages: an echo of the ancient promise that God will help if people turn to him, and that praying for 'daily' bread is advice to take one step at a time (whilst a long term objective is important, on the way towards it, it's one step at a time) and to trust in something out of our immediate control, and round the corner of what we can see.

4. 'Forgiveness'—ours and God's. Here Luke repeats Mark's earlier passing reference, that if *we* can forgive and make peace, then, and only then, will the great creating God love us and give us his peace.

5. 'Do not bring us to the time of trial' is a reminder of the 1st century plea to be strengthened against powers stronger than those of the individual. Other ancient translations illustrate this with, 'deliver us from temptation and rescue us from evil' (or, 'the evil one').

III. God's Special Concerns

The Beatitudes: Thirdly, standing beside these attributed words of Jesus is another set of text, where Luke declares Jesus' claim that God has special concerns for certain people. It's worth the modern reader thinking through these carefully to see if they should inform our own attitudes today. The sayings comprise eight statements in Matthew's gospel, of which four also are used by Luke, with some added comments. They appear to echo earlier words of Jesus adapted to meet the situations of Christians in the seventies of the 1st century.[4]

Each statement contains an assertion, and a consequence. God blesses (gives special favour to):

1. The poor in spirit; for theirs is the kingdom of Heaven.
2. Those who mourn; for they will be comforted.
3. Meek people; for they will inherit the earth.
4. Those who thirst for righteousness; for they will be filled.
5. The merciful; for they will be shown mercy.
6. The pure in heart; for they will see God.
7. Peacemakers; for they will be called children of God.
8. Those who are persecuted for righteousness' sake; for theirs is the kingdom of heaven.

The emphasis is on those who suffer from disadvantages and persecution in 1st century society, and presents God as an active force for social justice. The wrongs of today can be set right in the heaven of tomorrow. But just a moment! Is that the whole meaning of these statements? What about the references to, 'The Kingdom of Heaven'? These tie in with an oft used phrase of the historical Jesus of the thirties, that God's compensations occur not in the future, but within humans who seek him now, in the present. This changes the meaning of the beatitudes, and becomes a driving force of the New Testament. It is covered in the next section.

IV. WISDOM FROM THE FIRST CENTURY

The Kingdom of God: A fourth theme for the present is a mysterious phrase which bounces off modern ears without troubling them with a meaning. But it is worth stopping, pausing to probe into it, for it's another signpost for modern journeyers.

The phrase appears over fifty times in the recorded sayings of Jesus, and is clearly his major theme for his audiences in the thirties. It was based on earlier Jewish traditions that God was King of his people, a common theme in other ancient theocracies; and one which is still to be found in societies today, notably in Iran, parts of Egypt, North Korea and the IS zones of Iraq/Syria.

The phrase can be made accessible to modern minds in four main stages:

1. It has to be realised that no adequate translation of the phrase exists in most modern languages. There are differences in some of the ancient texts, and in English the Kingdom of God has been re-presented as, 'The Kingship of God' or, 'The Kingdom of Heaven' along with other implications of a promised land, a city of God, a New Jerusalem, a paradise—all of which fail to unwrap its mystery. This caused a gathering of protestant American Scholars in the 1990s in desperation to try, 'God's imperial realm', a noble attempt which unwrapped nothing of the mystery for European minds.

2. Stepping past the difficulty of translation, the phrase is clearly about a state of human living where Gods' will works within human minds and bodies, and drives them forward, not following their own priorities, but His.

3. It also includes a hope, born in Isaiah 700 years before Jesus lived, that God would finally intervene at the end of the age, and establish his will clearly on earth.[5.] This is harder for modern Western minds to grasp. It had developed into a continuing theme in Jewish belief that history was divided into segments of time in which the main religious events were repeated. The strength

of each revelation was thought to grow with each succeeding age. By the time of the 1st century, there was a strong anticipation of an imminent intervention and a forthcoming judgement. John the Baptist, Jesus and other charismatic figures all stood in this tradition, and the compelling thought of a coming return of Christ, the son of God, drove the extraordinary expansion of the movement in its early years. Subsequently the belief in this form of eschatology—a final victorious return, and the establishment of a kingdom—has been refreshed in later generations. However, nothing has happened to fulfil any of this.

4. But then these three steps all see the Kingdom as something beyond our sight in the future. I suggest that this was not the actual message of the original Jesus. This remains there in the gospels all the time, where Mark starts his gospel with Jesus coming from obscurity into the public gaze by proclaiming, 'The time is fulfilled, and the Kingdom of God has come near. Repent and believe in the good news.'[6]

The good news had an unmistakable political dimension (a fact not unnoticed by the ruling Jewish establishment). Whereas other movement leaders had encouraged followers to help build a situation suitable for God to intervene and set things right in Heavenly justice, Jesus was different. The kingdom of God was declared by him to be in the present, and to be engaged and strengthened by his followers, as they grew in understanding that he was the direct agent of God's purpose. As his followers worked this through after his execution, they came to see God as the real source of social justice in a society where there was little to be found, due to greed, lust for power and possessions of the people who controlled it. As the Lord's Prayer put it, 'Your Kingdom come, Your will be done.' (But where?) '...on earth as it is in Heaven.' God emerges from this scrutiny as a source of creative love and care, deeply involved in suffering—the biggest cause of which is social injustice. As Marcus Borg wrote in 2003, 'What would it mean for us all to take this seriously?'[7]

From this re-shaped view of the kingdom, a compassionate, testifiable form of Christianity emerges which can add value to our own analysis of our present and future potential. We can do this, each one of us, through a serious re-thinking and modern interpretation of the ancient texts. To begin to engage with this vision of the age-old phrase is to move out to explore another word with a modern interpretation: 'transformation'— the subject of a section in Chapter Eight.

V. WISDOM FROM THE MORE REMOTE PAST

Isaiah—A Highway to God: During this brief journey through the Holy Writings of the Christians, we have seen how they grew out of the long tradition of Jewish religion, stretching back through the mists, nearly to the dawn of recorded history. One reason why they endured through time, was their continuing relevance to each age which had inherited them. Their expression of a great bond—a covenant—between the Jews and their God, formed the total background to the faith of the early followers of Jesus, who came to see his message as a new bond with their guiding God in the 1st century—a New Covenant. Later this came to be recorded as the Old and New Testaments of the Christian Bible. The links back to their roots, which are everywhere to be seen in the Letters, Gospels and other documents produced in the 1st century, and I would like to add to these one more example of this wisdom—one further item to our list of useful values to give us direction in times to come.

To do this we turn to the book of Isaiah which is a collective name given to a number of scrolls brought together in the 5th century BCE, but drawing their contents from many earlier sources; some of them passed down by word of mouth from father to son. In this collection there were two prominent themes. One was a plea to the people of Judah to keep close to their God, who would reveal his worldwide Kingship in their country. The other was how this revelation could be projected from Jerusalem outwards to the whole of the known world. One example of these themes is the imagery of a Way, or a highway, linking the faithful and coming *from* God. But elsewhere in the books we see this as a highway *to* God.

This first theme appears in a section where the people have suffered for their errors, and Yahweh sends a message of encouragement, centuries later given prominence in Northern Europe by its inclusion in Handel's Messiah and other music of the re-formed churches:

'Comfort, Oh comfort my people... in the wilderness prepare the way of the Lord... make straight in the desert a highway for our God.'[8]

A glance at a map of 8th century Judah gives a colourful background to these words. There was Judah, the home of the Jews, in a narrow corridor of land through which the African, Egyptian cultures to the south, and the Assyrian, European lands to the north used as a link for trade, for military purposes, and for cultural exchanges. Another glance will reveal the problem. Save for a few settlements where people clung to a precarious and difficult life-style along a narrow strip of the coast and a minute valley following the River Jordan about thirty miles inland, everything else was desert. And in that situation, the vision of the writer was that the people could build a highway, and that God could use it to guide them and save them from their survival struggles.

The concept of a highway appears in other places in Isaiah, and a two-fold significance emerges. This highway—this link—was for the people to use, but was also for God to use, echoing again the Covenant between them. Since this theme, among others, has provided inspiration to subsequent ages, what does it suggest to enquiring people today? Can it provide any encouragement to people facing the enormous issues of the 21st century? I think it does, and that its current relevance is revealed in three ways:

1. A highway was used for travelling, and all of us are travelling, in a different sense, along our journey of life. But all journeys have a purpose and an ultimate destination. A call from the silent past to us, is to pause to decide the purpose of the remaining times left to us in our own life journeys, and to ponder where it's all going to end up. The next chapter will look at some of the alternatives including the unmentionable one—our own decease.

2. This sort of journey is undertaken by us as ordinary individuals, but how do we know the way? (Something that was a real problem to the map-less illiterate age which produced the legend in the first place). When the Christians re-stated the Isaiah theme, Jesus the Christ replaced Abraham, Moses, Isaiah and Ezekiel of earlier years, as the guide on life's journey. Writing eight centuries later, Matthew puts into the prophet John's mouth the

same words of Isaiah, and follows them with the account of Jesus being the light for people who lived in darkness. Could it be that on our own journeys, the example of Jesus is a daily continuing light for the enquirer, and that the ancient writings are like a map for us to follow—whether printed or electronic—perhaps a sort of sat nav?

3. The journeys of past ages have frequently been made in groups—for practical reason of security and support. The groups on medieval pilgrimages were typical examples, as mirrored in the 14th century by Chaucer in his lively Canterbury Tales. Our life journey too, can benefit enormously by being undertaken in groups. But the spirit of much of our age inhibits this—with exploration and entertainment becoming so personal to the holder of an iPad or a mobile phone. A tip from the past is to enrich ourselves by good company throughout our journey. It lessens the load, and is worth bucking the current trend of individualism, to re-discover.

I always enjoy Isaiah, and even more so the other examples we've read of the wisdom of the past being stored in the Christian Holy Writings. But there is another source of wisdom and truth stored up in the Christian tradition, the huge body of teaching and good living advice built up in the Churches and passed down from age to age.

The Legacy of the Christian Churches, Past and Present

This enormous store of guidance has helped successive ages to face their own problems, refreshed by the experience of those who have gone before. Like the Holy Writings, this inheritance has to be handled sensitively and with intelligent caution. Many a ruler has used the doctrine of the Church for his or her own ends, and many have used it to justify their own military adventures and repressions of other people's freedom. The church has used its doctrine for the exploitations of others, for slavery, for sexual promiscuity, for the suppression of women and the persecution of minorities who won't or can't conform.

That's the trouble of having humans in the church! But there is another value entirely in this body of teaching which, like the cleansing of the people mentioned by Isaiah, and the repentance theme of John and Jesus, can be sifted of its problems to reveal genuine treasures of truth and wisdom within them.

As previous sections have explained, the wisdom of the ages has been given added value by its refreshment from current ages. It is also true that the earliest words and examples have themselves gained added value from their use in worship, meditation and prayer through 2,000 years. So the lists and codes of living the Christian life which originated in the 1st century, have been stored faithfully in Christian doctrine and have influenced much of the growth of Western Civilisation. They are here and alive to guide us today. As the 1st century writer of Hebrews wrote:

> 'The word of God is living and active, shaper than any two edged sword, piercing until it divides soul from spirit, bone from marrow. It is able to judge the thoughts and intentions of the heart.'

The next chapter will seek to show how the Christian roots of our society still feed thoughts today which can indeed help add extra value to decisions we all need to take about where we are going in future.

NOTES TO CHAPTER SEVEN

1. The number of Biblical Books is rather more complicated than this. The figure quoted is that established by protestant churches. They excluded a number of other books accepted by the Roman Catholic Church, and other ancient churches had their own canons (collections) with more variations. Additionally, further documents not known at the time of the main agreement on a Biblical Canon in the 5th century were discovered in the 19th and 20th centuries. The Dead Sea Scrolls form an example (see *Bibliography.*)

2. *The Great Commandments* can be found in Mark 12:28–34, Matthew 22:34–40, and Luke 10:25–28. The quotations from the older Jewish scriptures can be found in Deuteronomy 6: 4–5 and Leviticus 19:18.

3. *The Lord's Prayer* appears in Mark 11:25–26, Luke 12:1–4 (passing reference) and in Matthew 6:9–13. Also to be found in two contemporary non Canonical sources, Didache 8:1–2 and the Gospel of the Nazareans.

4. *The Beatitudes* can be found in Matthew 5:3–12 and Luke 6:20–23.

5. The closing 14 chapters of Isaiah, starting from 52:7 record the vision and hopes of a suffering servant of God, that a greater age would dawn in which peace, humility and justice would prevail, with the enemies of this process being eventually over-ruled and crushed. This theme was developed by other writers in following ages, and fuelled the 1st century revolts against the oppressive military occupation of the Romans a generation later. The Christians echoed this in their writings of the same period.

6. *'The Kingdom of God has come near.'* Mark 1:15.

7. See Marcus Borg, *The Heart of Christianity* (2003), *Chapter 11: Heart and Home: Being a Christian in an age of pluralism.*

8. *The highway to God.* Isaiah 40:3. See also Isaiah 35:8.

CHAPTER EIGHT

CHRISTIANITY: WHAT ARE THE 'EXTRAS' IT OFFERS?

When I began an earlier chapter with a story about pebbles on a beach, I didn't quite give you a full account. I'd better try again. The story really begins back at the dawn of the 20th century, with a Victorian child beside the sea playing with the pebbles. The sparkling colours and shapes filled her mind, and she didn't notice the huge wave bearing down on her. Suddenly to the horror of her watching family, she was swept away and disappeared. The terror of such a moment is hard to imagine, but was cut short by the instinctive response of her father who plunged into the swirling tide, flung out a hand, found her, and pulled the child out through the breakers by her long ginger hair. Pumped out and warmed up, all was soon well ... except that the child was traumatised by this throughout ninety long years—the whole of her life. I know that—for she was my mother.

So imagine what flashed through my mind a generation later, when the same thing, by complete coincidence, happened again. Although carrying an inborn fear of the tide, I too had to plunge in to pick up a child who'd been swept off her feet by a freak wave. But both she and I benefitted from that, because I conquered my programmed fear of water, and she who now snorkels around fearlessly with her own children, completely able to cope with that environment.

There are two morals in that story: one is about how we feel if something snatches us away from our normal security, and the other describes how succeeding generations can overcome the fears of those who went before them. The first of these is a daunting anxiety. We, like the pebbles on the beach, are picked up by the tide, and swirled along. Threats and challenges we have never known before pressurise us, and we sense a drowning mix of things we can't control; environmental change, over-population, mass migrations, and a shortage of food and natural resources, to say nothing of the greed of humankind. We clearly need all the help we can get, and will clutch at any things that might steady us. Some of these have been outlined in earlier chapters suggesting how experience of those who have gone before us, can give us the tools and techniques to help us overcome our own problems. Science, pragmatic knowledge, technical improvements, growth of knowledge; all these can all help us take the right decision to lead us towards fulfilling future lives.

But since we do need all the help we can get, is it worth looking at the other source of human knowledge—that of religious experience, to see if that offers anything extra? Something extra to clutch? That is why the last two chapters have listed some of the themes that religious approaches to wisdom have embedded in our civilisation. Because of its huge contribution, some guiding points from Christianity have been illustrated, and this chapter now summaries those, and incorporates them into three particular themes which are changing rapidly today alongside the flowing tide of late modernity. They are the nature and effect of prayer, our attitudes to death and beyond, and our ability to achieve what we will call 'transformation'. Taken together, they form six 'extras' that Christianity can inject into our futures.

Extra Number 1: The Resources of Stored Wisdom

What are the extras which come from Holy Writings and Church teachings passed down through the centuries? Chapter Seven gave some examples of sayings of the earthly Jesus, but later generations put these words into writing to explain their relevance to their own age. It's therefore not surprising that the writings of these early followers list attitudes and attributes that enabled them to cope with the specific pressures of their own times. In Paul's letter to new Christians in Galatia for instance, he contrasts a life without God with one that has become filled with His values.[1]

He lists the human weaknesses he saw around him in the world of the 1st century as sexual waywardness, idolatry, strife, anger, quarrels, and drunkenness among other things, saying that these inevitably separate people from God. However, there is another list of things that do not; 'By contrast, the fruit of the Spirit is love, joy, peace, patience, kindness, generosity, faithfulness, gentleness, and self-control. If we live by the Spirit, let us also be guided by the Spirit'. He adds that building up the good qualities of others and bearing each other's burdens are the good ways to live, and warns that harsh attitudes to other people should be avoided.

'All must test their own work; then that work, rather than their neighbour's work, will become a cause for pride... God is not mocked, for you reap whatever you sow.'

Again, when Paul wrote to followers in Colossus and Laodicea he touched on a further dimension, that these treasures of Christ's wisdom were not confined to the original circle of followers in the thirties, but were transferable to people in later ages. From prison he wrote:

'I want you to know how much I am struggling for you, and for those in Laodicea, and for all who have not seen me face to face. I want their hearts to be encouraged and united in love, so that they may have all the riches of assured understanding and have the knowledge of God's mystery that is Christ himself, in whom are hidden all the treasures of wisdom and knowledge.'[2]

How do we transport the meaning of these thoughts from a remote generation to ours? After all they have passed the test of time and inspired successive subsequent ages. How can we recapture the original thrust and use it for our own guidance? Here we have to tread carefully, and intelligent study of history is essential (and also of the actual texts, which have their own field of study—hermeneutics). But briefly treading through the history of the resources of stored Wisdom, most of the Holy Writings had taken shape by the end of the 1st century CE, and over-lapped with the spread of the new religion to other parts of the known world, and also to people outside the original culture of the Jewish faith. A distinctive cluster of Christian communities had begun to spread, using the writings for worship, for teaching and for personal meditation, and a body of new, revised principles for Christian living had begun to take shape. The second foundation of wisdom was starting to appear—the doctrine of the Church.

The detailed history of this enormous spread of advice and inspiration is outside the scope of what I write here, but it spread quickly during the following centuries throughout the two main traditions of the Western and Orthodox (Eastern) churches. An important thing to note is that as each generation studied the early writings, so they gained added value in the minds of later followers, and refreshed and reinterpreted their meaning. The great outpouring of human energy at the end of the medieval period in the west, had at its heart revived understanding of this Christian message, and was fed by it. Its very freedom of course soon shook the institutional church, and both the inspiration and the rejection are parts of what we have to understand today as our heritage. But what further 'extras' does modern faith now provide for us in the secular world? Read on.

Extra Number 2: Prayer—Its Nature and Influence

The phenomenon of prayer is as old as recorded history, and probably much older. Its origins lie in ancient man's feelings of insecurity, and his sense that something greater around him could be called upon to help. At its most basic level it is a desperate plea for protection and survival, and remains in us all today as our last hope when in peril. Throughout history it has been a central theme of religious faiths, and its range has spread far beyond the basic plea for help, extending to adoration, invocation, giving thanks, expressing failures and seeking forgiveness. Prayer is undertaken silently, or vocally, and by individuals or groups. It is in the heart of the hermit seeking personal elevation to the closeness of God, and in the cries of thousands of followers at a mass rally. But what is it? Can it really make a difference? So many philosophers have asked these questions, so many theologians and scientists have argued them through, that we may safely conclude that nothing can be proved. Yet I would like to suggest that in working out its relevance for the 21st century, there are three themes to ponder.

1. Prayer has inspired, maximised and refreshed people for centuries, and the records of the ages are its testimony. While some societal practices were deemed ineffective and died out, prayer is one of those that has

continued. Something still draws people in trouble to pray.

2. It appears to be able to introduce reactions among users, not capable of being shaped entirely by themselves. The person who forgives the murderer of their loved ones, produces a flow of goodness perhaps greater than their own. The person who prays or finds new approaches revealed in dreams or life changing vision may well exhibit some linking between human and external consciousness—in short, God. This does not satisfy scientific or philosophical requirements of truth, but I record it here because the manifestations remain.

3. Modern science has started to reveal the astonishing insights of new physics and biology outlined in the last sections of Chapter Five. What was a theory, becomes a model, and approaches scientific definition.

If it starts to be credible to see life forms relying on personal information banks, stores of necessary energy outside their own visible bodies, what is the nature of the energy that flows between them? If a physical movement in one part of the cosmos causes a parallel one elsewhere, what actually flows between them to achieve this?

The historian and the theologian is now able to look at this new growth of learning and set into its matrices the thoughts that 1st century man was trying to express, when he described the power of prayer, and the consequences of the slightest change of consciousness or activity in every individual. Jesus appeared to be aware that God saw everything that humans did, even if only in their thoughts. Speaking to his critics, he said that while committing adultery was against the strict Jewish laws, a higher standard was required of his followers, for, 'Everyone who looks at a woman with lust, has already committed adultery with her in his heart'.[3]

So the energy within prayer will become easier to define, but this leaves behind one further question. Can a religious believer's prayer change the course of future events? If it cannot, has it any real value? If it can, then is securing special advantages for the skilful supplicant a form of dark magic, demonstrably unjust.

I open this up to wider understanding through the pages of an Edwardian journalist's book on Watermills.[4] A surprising but effective choice. As he described how they worked, the analogy with the effect of prayer is there to be seen.

In the table below, the watermill mechanics are in the left two columns, and the religious analogies in the right one—including the power of prayer.

WATERMILL MECHANICS		RELIGIOUS ANALOGIES
A river flows from high ground down to the sea.	Moving water cuts out its shape as it flows along.	The power of a creating God surges though time.
People build a watermill beside it.	Some of the water is diverted through it.	A tiny fraction of God's power flows through man's creation, in parallel with his own purpose. Prayers occur.
The wheel turns, and energy is produced.	The energy grinds corn or generates electricity.	Prayer produces results in energy released, work done, and help for others.
The diverted flow curves back towards the river.	The water flow re-joins the main torrent.	Human effort has increased God's work.
The river flows onwards into the sea	The sea is vast, and stretches over the horizon.	The energy of God swirls on to times and places we cannot see or understand.

Note that in this analogy, there is no magic involved or favouritism sought or granted. It is instead a process of people linking to the great creator, joining in as co-creators, and using the power of God—shall we be more specific and say 'love'—to further his eternal purpose as it surges onwards to the great beyond, which we can neither see, nor imagine. Prayer. It is an extra contribution to the world we hope to build.

EXTRA NUMBER 3: A REVISED ATTITUDE TO DEATH

HOW SOCIETIES CHANGE THEIR ATTITUDES

Our attitude to death is not just an academic exercise. Far from it. It includes our own demise and is difficult for our current age to contemplate. So an important starting point for us is to realise that different ages and different societies which have gone before our own, have understood death in different ways from ours. Is that because we now know so much more and can safely dismiss the past? Or might it be that older visions might have something to offer to us with after all? Many early societies needed to believe that death was not the end. For some there was the hope of reincarnation here, perhaps in human, perhaps in some other life form. For others there was the prospect of a journey to another world finding the company of a God, or the Gods. For others yet, there was the expectation of human bones being restored to life, with a deity pronouncing the end of time, and a general reckoning and judgement, with the good being raised to eternal life, and the evil ones sinking to eternal damnation. The Old Testament prophet Ezekiel dramatically uses this image in the 6th century BCE writing symbolically of the valley of dry bones within which the people of Yahweh could rise above their declining lifestyle and be raised back to life.

This wide range of views, many of them alien to our own, springs from the life experiences of those who held them. Where oppression and corrupt power prevailed, we see the belief in a balancing process of justice and judgement existing on the other side of death. It just had to come right eventually, that hope alone justified the sufferings of the present. In feudal societies throughout Europe the sufferings of the poor and entrapped, were seen as trials which were capable of post death remedies, through the God of love and justice.

In modern times we have seen the power of the Orthodox and Roman Catholic churches behind different parts of the iron curtain encouraging life beyond death as a realistic prospect, while in the more settled and comfortable west, this understanding has quickly faded away. In the caliphates of Iraq and Syria, a strong belief has developed in the power of Allah rewarding those who have suffered under past corruption and faithlessness. A prospect of eternal justice grew, guided by those who remained faithful in life, or heroic in death through martyrdom. Vivid historical examples support all these views.

Yet in the west today, despite all our pressures and fears, there exists a more general stability and affluence that has come from the slow growth of democratic freedoms and new technologies over many centuries. Consequently, our need for peace and justice after death diminishes. With two results.

The first result is that we can work towards these things now, and turn away from thinking about death—because we increasingly see less of it as medicines develops and people live longer. We can make our own decisions and do that now, without waiting for a 'beyond'. Surprisingly there is a strong religious parallel. The mysterious Kingdom of God described in the 1st century writings, is not another place situated somewhere beyond death, it is, as Jesus came to proclaim, available here and now. It is for people to work for now, to set up a context serving God's purposes, not the greedier ones of humankind.

The second result is both good and bad. Good because longer lives and less pain are amazing privileges for people today; yet bad as it makes us turn away from the longer term issues that yesterday's societies were able to take on board quite naturally. Why did they build huge cathedrals, greater than anything the worshippers could ever have needed for their own use? God was vast and beyond, and they were reminded that they were minute specks as they contemplated him. There was something beyond their immediate view. They would not forget it. Yet for us today, beyond all our public discussions and exchanges of views on death and its consequences, there remains a zone of silence. We may discuss the issues around prolonging life or choosing death, but after that... what? Nothing. That is the contrast with so much of previous human expression. Religious perspectives do have something extra to say, but its exponents need to find the right vocabulary quickly.

The mind-sets of the medieval cathedral builders, are in fact being paralleled by discoveries of new science which reveal the same attitudes, but with even more power. New knowledge of time and space brings us back to astonishment and humility. For the believers in religious truth to understand this and remodel their beliefs to take it all in—that too becomes another extra for the future hope of the modern global world.

The Great Beyond—Is There Life After Death?

> *'It seems to me most strange that men should fear;*
> *Seeing that death, a necessary end,*
> *Will come when it will come.'*

Shakespeare expressed this pragmatic view in 1599,[5] yet it would serve today's secular world well. Death is clearly the end of each individual's life. Full stop.

But, with even more ancient roots there is another view typified by a biblical writer, Paul, fifteen hundred years earlier. He writes:[6]

> *For we know that if the earthly tent we live in is destroyed, we have a building from God, a house not made with hands, eternal in the heavens, for we walk by faith, not by sight... we would rather be away from the body, and at home with the Lord. For whether we are at home or away, we make it our aim to please Him.*

This conveys the earlier belief that in a sense they did not quite understand, death was not a full stop—something carried on and linked with God in the eternity of the great beyond.

For us in the 21st century, which of these views is believable? Having studied the life and growth of Western Civilisation, and its foundations in science and religion, it's my conclusion that the answer is; both— though they each need to be explained in terms of what we now are coming to understand. A new view needs to see these together, and form a bridge like this.

Foremost, it is clear that death of the physical body is final and not reversible. People do not physically have their deceased body cells re-made and do not rise again. But is that the end of their influence and presence among those who remain? The emerging answer appears to be 'no'. Some of the conclusions of new biology point to the information field of the deceased being accessible to those who draw similar data from theirs, and while they continue to do that, the presence and influence of the departed remains. Rupert Sheldrake was pointing out that forms of animal and plant life could acquire characteristics of their predecessors—with a dog being terrified of butchers because his father had been mistreated by one. So in this sense it is possible to see humans living on in their children's consciousness, through their shared access to particular fields of morphic energy. What is more, emerging science opens the way to verify the transmission of thoughts and attitudes to those we may have influenced as teachers, or friends. As we think this through, the extensions of these analogies continue to expand. Fragments of the consciousness of great writers, thinkers, composers, poets, remain alive in others while they continue their links with the originators' fields. So while the residual influence of most of us does not continue in this way for more than a couple of generations, some do, and these continuing energies inspire followers, fortify societies and make them (and us) what we are.

The religious are not surprised by this, and see the great figures of their own traditions remaining as live influences speaking from the past, but being very much part of the present. It is a different way of 'knowing' inherited truths, but there to be seen.

Perhaps that is the mechanism by which Jesus is seen as alive by millions of Christians today, and perhaps that is yet another extra which religion has to offer to supplement science. Writing in the 1990s, Bishop Spong added two other pieces of background to this theme.[7] One was the huge increase of human awareness which had already grown considerably by that time, underlining our inability to comprehend or measure the vastness of life, 'There's always another level stretching away into the distance'. This meant that he could not rule out a form of afterlife. His second piece of background was more tangible. It was the power of love which he observed in people's lives, which, 'enables being to emerge in each of us.... It is a journey into God without limits... when I have travelled then I believe I have touched that which is timeless eternal and real'.

Many more changes in religious and scientific thinking have occurred since these writings of the 1990s, to point to something in each of us which is capable of enduring—capable of transmission to the hearts and minds of others, and forming a further posthumous chapter of our personal contributions. In a section of a book asking what extras the religious view can contribute to the future, this is the biggest so far.

Of course any fusion of new energy with old visions calls for changed and expanded views, and high on the list of change is the need to re-focus on the appearance of Jesus and Christianity on the scene of world history.

In 2010, Dominic Crossan, joined with Marcus Borg to visit and study surviving evidence of early Christendom. Their researches revealed that during the five centuries when the Western church was taking shape, the Eastern tradition held on to its older view that there was a remarkable transformation in the minds of the early followers, who saw Jesus as alive in their hearts. He was pulling them forward to leave the graves (the situations) where their own human weaknesses had got buried. New energy and confidence filled their lives. The actual physical resurrection of Jesus was not at the front of this argument. But for the western branch, Rome, fighting against the claims that the Emperor was God, it was. So our later civilisation became focussed on one man's resurrection, where the other major part of Christendom gasped with amazement at a transformation.

The refocussing is called for, not to deny resurrection, but to understand it afresh within the whole amazing context of human transformation—something available for us today, unlike physical resurrection!

We will ponder this soon as Extra Number 5, but by way of preparing for that we turn now to some further information about scientific probing of the ways in which we sense the beyond.

EXTRA NUMBER 4: SENSING THE BEYOND

In one sense, there is nothing new here. Humans have frequently demonstrated an advanced sense of the beyond, as they drew nearer to it. Wisdom on the deathbed, nobility on the prisoners' death row, and statements of the martyrs have all shown a heightened awareness of a presence greater than theirs, as it started to surround them.

But in another sense we now become aware that something has changed, with the growth of new information, not available to past ages. A huge increase of popular web access to new developments, developing medical knowledge and enhanced knowledge in other sciences are the three factors behind this. The story grows.

One further example of the consequences of the spreading internet has been the increased awareness of ways in which thousands of people feel they sense the beyond. For instance, the data about Near Death Experiences has significantly increased and claims of these experiences have gathered momentum (and of course sometimes been denied or discredited—two sides of the information revolution).

However, the growth of medical knowledge is different, because it is more measurable. It can throw new light on a previous query about what happens to our energy field after we die. Can this live on? Semyon Kirlian, a Russian quantum physicist and academic who died in 1978, had developed a technique of detecting energy passing through human tissue by creating a halo of light that could be captured on film. This process has more recently been modernised by another Russian physicist, Konstantin Korotkov, using television and computer equipment not available a generation earlier.

He demonstrates that the light captured was the energy field of a human being, and in a series of experiments in the late 1990s showed that for two days after death there was no significant change in the radiation from the body, declining thereafter at a rate which reflected the lifestyle and nature of death of the deceased. He concluded that the measured light was living after the body had physically died and only subsequently disappeared on time-scales which differed according to the individuals concerned.

This form of creating new scientific models of course has many detractors, and a debate currently continues, but its interest for the point of us here, is to see how this measured conclusion fits in with the more primitive imagery of 1st century thinkers with the orthodox Jews believing that what they called the soul lasted for three days after death. This was exampled by their expectation that after Jesus' execution there would be a waiting period of three days and three nights before resurrection and transformation became passed on to others.

Another benefit of increased medical knowledge is the greater information now available about life expectancy following major illness. More accurate statistical probabilities about longevity means that patients now have an extended period for adjustment, planning, thinking and sharing with others. This is important to those contemplating their demise, and a new phenomenon. Whereas an increased awareness-period anticipating the beyond used to be undefined or brief, this can now be extended for months or even longer. During that time people change their attitudes, become more sensitive to others around them, more receptive of the idea of something of them continuing, and less inclined to accept that at the moment of death everything stops. I have personally seen a strengthening of moral values in some folk, and a developing sense of reconciliation and peace in others. I cannot measure these things, but feel they represent new opportunities to grow during these periods. For people of religious faith, the feelings are differently expressed and form a feeling of growing proximity to God, and the start of a personal period of adaption and change—transformation. We take a further look at that next.

EXTRA NUMBER 5: THE HOPE OF TRANSFORMATION

What does it mean to have such a hope? Why should we need one? And if we do, how do we set about getting one? Does it mean anything? These questions take us back to understandings of the early Christians in the 1st century. Their earliest writings suggest that it means a change in lifestyle from worldly ways to others which are more pleasing to God. The transformed had minds which became changed—refocussed—in ways that altered what they said and what they did. In powerful words addressed to the followers at Corinth, Paul wrote of a new way in which God could be perceived, by people with, 'unveiled faces seeing the glory of the Lord, as though reflected in a mirror'.[8] They were 'being transformed into the same image, from one degree of glory to another... this comes from the Lord, the Spirit'.

The strong feelings of transformation, had thoroughly practical consequences. They turned timid men into brave ones, followers into leaders, and the people on the fringes of communities into those who were valued and accepted nearer to the centre. In the centuries which followed, the Christian movement spread though all levels of society in and around the Roman Empire, and effectively proved that the Jewish old guard were right to fear that these people were, 'turning the world upside down'.

All of this in turn reflected what the early disciples had seen in Jesus, transformed from manhood into Godliness, by inner development within his mind, affecting his actions, and eventually bringing him to trial and execution. But the theme continued, and was cherished by the early Church, with the accounts of post death appearances to selected followers in specific locations—something which the Jewish followers were anxious to explain as repetitions within a cycle of history, of earlier revelations of God's nature through Abraham, Moses, Elijah and many others.

Returning to our present, it's interesting to see the parallels between the implied energy transfer from the external force (God) to the recipient humans, and the analogies of energy transfer being modelled in new physics and biology. Lynne MacTaggart, whose work we referred to in Chapter Five, has followed this subject for two decades and draws together the experiences of other scientists in developing the idea of a sea of energy, a cobweb of energy exchange, reconciling mind with matter, classical science with quantum physics, and science with religion.[9]

EXTRA NUMBER 6: RESTATED VALUES— THE FOUNDATIONS OF THE WEST.

In the countries of Western Europe, our values clearly come from a number of sources; the philosophy and economics of the last five centuries, from new knowledge throughout the sciences, from the hi-tec revolution, from the creative events of the post-world war era, with United Nations agreements on state, societal and individual rights, and also from the growth of international law. All of these form the basis of secular democracy, as we understand it in the west. However, all of these also originally grew from foundations in the Christian movement. Within the first five centuries CE, this continued to evolve and re-focus, still placing further roots through medieval ages, until becoming much eclipsed by the driving force of secular democracy taking shape in Modern times.

Yet now, as science and religion begin to find a sense of new togetherness, the need for these two areas of human knowledge to respect each other and converge is nowhere more apparent than in the concepts of energy flow and of human transformation. Are our brains not just local terminals, but transducers linking out to other sources far beyond our grasp of time or space? Is 'God' the everywhere field of moving energy everywhere in the universe.

These profound themes all help us to harvest the wisdom of the past, contemplate the huge issues of the present, and point us onwards to forge a new set of objectives, both personal and global, to guide us out into the future. That is the role of our final two chapters, to which we are now equipped to turn.

————————

NOTES TO CHAPTER EIGHT

1. *Paul's letter to new Christians*, Galatians 5:13 onwards.

2. *Paul in prison*, Colossians 2:1-3.

3. *Adultery*. Matthew 5:38.

4. Hervey Benham, *Some Essex Watermills* (Essex County Council Newspapers Ltd, 1976).

5. Shakespeare, *Julius Caesar*, Act 2, Scene 2 (1599).

6. Paul, writing to new Christians in Corinth within a generation of the life of Jesus, 2 Corinthians 5.

7. J. S. Spong, *Why Christianity must Change or Die* (New York: HarperCollins, 1998).

John Shelby Spong (b.1931) is a retired Bishop, author and international lecturer. A committed Christain he is regarded as the champion of an inclusive faith who makes contemporary theology accessible to the layperson (see *Bibliography*).

8. Paul, (as in *Note 6* above) 2 Corinthians 3:18.

9. See Lynne McTaggart, *The Field: The Quest for the Secret Force of the Universe*, (Harper Perennial, 2008).

website: www.lynnemctaggart.com

CHAPTER NINE
CONCLUSIONS

So now we pause to consider where we have all come from, where we are positioned at present, and where are we going as we move into the future?

What is our toolkit for repairing things past and present, and for building a better tomorrow?

We start with some private thinking about our own intentions. I'm going to write mine down (but not here) as bullet points for my eyes alone to see—you may wish to do the same.

Five Personal Questions

I'd like to suggest that we turn to pencil and paper, and make a list of some key themes, adding any further ones which apply to our own personal situations.

- How long have I got?
- Realistically, what are my main talents to offer?
- What issues really matter to me?
- How can I further these intentions?
- What could the possible effects be of doing these things?

How Long Have I Got?

Asking how long you have, is a practical starting point. It requires us to make a rough guess, as we peer into the future, as to how long—how many years—do we anticipate having available to us of reasonable health of body and mind to carry out our life intentions. It's a pragmatic question—nothing to be frightened of. Let us include foreseeable financial changes, and developments for better or worse in our social contexts—family, friends and spheres of work. What length of an active period does this give us, before we have to retreat into ourselves just to keep going?

What are My Main Talents?

This provides a time frame in which we can fit a major piece of future thinking. What are my main talents which I can grow and use during these years? Although I've said it's a private process, if it's possible to ask someone whose judgement we trust to tell us what they can see about us, that would be good. None of us really knows what positive influences we can deploy without knowing how we affect other people. Two or three bullet points could be enough, to mix metaphors, rather like hooks on which to hang our future visions. An old farmer I knew always had his supply of useful hooks ready beside him. They were aids readily available to him if an outside challenge suddenly arose. 'It's 'anging on an 'ook in the 'all' he would boom and reach up for the oilskins, torch and medical kit hanging there, before stumping off to another cow or sheep crisis. We too need to know where our facts are, and what our attitudes are, before critical issues call upon them. Like the old boy, we need to be ready, with the right responses at the front of our minds.

WHAT REALLY MATTERS TO ME?

The third step to take is to consider what issues really matter to us. Whereas the first two asked us to think of our positives—our potential to make contributions for good, for us and for others, this third one points us towards surrounding issues which may have a quite different feel about them. They will include things that threaten us and, frustrate us. Things that we fear we cannot change. Once again it's the spectre of us as pebbles being swirled along in strong currents. They may involve our relationships, our financial and social security, and they will in all probability not be capable of solution in a short period—if at all.

But what do all these steps really amount to? What are the three or four main things that we would like to change, or see changed in our surroundings during our remaining active years?

How Can I Further these Intentions?

There is an old training adage within Personal Development courses of giving a plan an immediate starting point. Starting: that's important for us. All around us we see ordinary people making a start to their plans; the mother who sets up a charity after her daughter's death to prevent others suffering the same fate; the plumber who retrained as a doctor after seeing the support a GP gave to his sick wife (he later became a specialist surgeon); or the footballer who became a leading LGBT[1] activist after his own experiences of prejudice and ridicule a decade earlier. Our written plan is not enough. It has to be turned into action—and a selected starting point is crucial. Often this can be helped through the companionship of a friend who thinks the same way, and yet another encouragement is the enormous help publicity can bring in its wake—often at a much later stage, when growth suddenly bursts out.

It's a remarkable feature of our age that ordinary people now have the means available to them of accomplishing extraordinary achievements affecting other people who they don't know and never will. We are leaving the age where our influence dies with us, and entering an expanded age where the mark we make on our world can far outlive our own more limited origins. Whereas in the past only great people were remembered much after their generation, many more can now change things tomorrow by what they have started today. That is all thanks to the digital revolution, to increased knowledge, and to the huge growth of global society.

WHAT COULD THE POSSIBLE EFFECTS BE OF DOING THESE THINGS?

So to our fifth and last step. What could the possible effects be of doing these things, in the field of personal activities in which we plan to involve ourselves? For us as individuals there are huge benefits. Our remaining active years suddenly grow a new purpose, with increased confidence, more valued friendships, connections to groups, and a sense of growing fulfilment.

Much of this sense of growth is provided by becoming more aware of the heritage of the past and of its enormous store of wisdom and know-how, there to guide us. As we become more sensitive to that, we start to see the approaches, ideas and attitudes that will not work, and more significantly, those that will. We will also feel the enthusiasm from doing things that fit in to our plan, and start to build it up. Refer to the plan regularly, and make sure that our latest actions and intentions still fit into it.

If our personal intentions are effective, then they will surprise us by vigorous growth. We may or may not think of making a mark on the world, of influencing it for good in our remaining time, but the fact remains that now it *is* possible to make a difference. Most people can pass on influence to those who follow, but this may not last more than a generation or two. For some with more to say, this can be extended, as evidenced by great lives in the past. Yet now, things have changed. These opportunities are made available to ordinary people— people like us, a thought which leads us to the final section of this chapter.

Applying these Intentions: Personally, Nationally, Globally.

Five-point personal plans are capable of influencing wider society in myriad ways. Similar groups of intentions come together, trends and patterns are formed, and changes start to occur across the whole of society. These may take the form of spontaneous reactions to outside pressures, but equally can be directed by the planned intentions of groups or individuals. All around us we sense these change mechanisms at work, and come to see the wider view that it's not just a process within a national culture, but part of a more far reaching expansion of relating to the global world. Our intentions affect other people far away, and theirs alter our ideas and opportunities also. So a personal plan or an individual movement can spread at the speed of an electronic signal to link with others worldwide. We can look back at our past heritage and draw conclusions, but now we have the challenge to launch into the future with new beginnings. The last chapter sets the background within which these can grow.

––––––––––––––

NOTE TO CHAPTER NINE

1. *LGBT.* Social rights movements campaigning on Lesbian, Gay, Bisexual and Transgender issues.

CHAPTER TEN

BEGINNINGS

'Flourish' and 'decline' are adjectives for every age with the pointer of history eventually moving to show the appropriate label. It's therefore hard to assess recent times in these terms, and still harder to judge the present. As we have seen, the start of the 21st century displays both, with declines of old moralities, and over-stretching of resources on the one hand, and the beneficial opportunities of new knowledge, hi-tec and globalism on the other. So our starting point for journeying into the future, has to take full account of both of these possible verdicts, and all the variations between them. But in today's shifting, changing, accelerating context there are two ways of thinking within which it will be possible to create a better future for humankind, in spite of all the pitfalls and difficulties. They provide possibilities of a new background of converging knowledge on which sounder future foundations can be laid.

RE-CONNECTING SCIENCE AND RELIGION

The first of these is to piece together the signs that some of the outer fringes of both science and religion are starting to converge. After centuries of each rejecting the claims and intrusions of the other, certain glimmers of similarity start to appear. The Newtonian model of the universe is replaced by the loop theories and gravitational fields of new physics, whilst the literal version of a living God creating everything in six days, is replaced by a new understanding of what the ancients were trying to express within the only available imagery of their own time. Some of the underlying structures of the two new positions are not all that far apart. So can the language of both be re-presented so that each can start to accommodate the other? The same possibilities of convergence can be seen in the new biology of morphic fields, and the physics of attracting energy from space—a form of a 'cosmic free lunch' with both probing ways of using minute energy particles.[1] Again, across the other side of the divide, liberal theologians ponder the nature, mechanics and effect of praying, with minute energy moving out from humanity to God across the unimaginable dimensions of space and time. Once again, the recent developments in minute particle physics are becoming paralleled by new theological thinking on the energy of Spirit, and the survival through time of the expressions of those who have gone before us. Each discipline can help the other by creating new models, testing them within their own collegia, and bringing together their revised understanding. Such a convergence would have been unthinkable a generation ago, and still remains a huge process with much to obstruct it—but I write these words because it is now possible to sense a beginning.

RESTORING THE SOURCES OF HUMAN KNOWLEDGE

The second background theme comes into play when one realises that throughout history Religion and Science have been the two receptacles in which human knowledge has been stored. There has been a long history of struggles between them, with seven centuries (in the west) of growing domination of scientific knowledge and the often embittered retreat from it of the Christian Churches. From this has flowered the spirit of secularism leaving the religious pursuit of ultimate truth to be reduced probably to less than 10% of the current UK population. This imbalance, seen through the eyes of history, emphatically reverses over 3,000 years of earlier experience.

As convergence begins to dawn, the main attempt to span this chasm must come from the religious constituency, because it is their vocabulary and imagery that have become untenable in the eyes of the huge majority. So can this be done? Clearly convergence is only possible if the religious claims to different ways of seeing truths appear to be credible, and to offer better prospects for the future. In earlier chapters, emphasis has been laid on the value of this view, because of the evidence that there is indeed extra value to be added from the world's major religions—refreshed and re-presented—which can enhance life journeys and provide for a better future for humanity.

In short, science is not much use when societies cherry-pick its benefits and expand wastefully without care for others. Conversely, Religion is not much use when it ignores modern knowledge and closes minds and hearts around yesterday's tenets. But if links grow, and one starts to be informed by the other, the whole basis of society can change for the better.

Building a New Future

One further question remains to be asked. Who are we building a new future for? If major changes in world issues could be made in a decade, then the answers could possibly be 'us', but realistically the time and scale of rebuilding our great civilisation stretches far beyond this.

So as the past has done before us, it is now our turn to build for and care for those who will follow us. For those with children, it is a natural instinct to want to provide for them, and for grandchildren too. That goes much deeper than bequeathing possessions, for it extends to passing on our specific knowledge of how to navigate through local and family problems, and to growing in others the gifts that have served us well. Even wider than that, there are opportunities far beyond our family circles to leave contributions we have created for the benefit of society more generally. Those who set up charities to rectify the shortcomings in their own generation would not wish them to die when they do. Those Victorian railway builders who left generous bridge spans for those who succeeded them, benefitted the people in our own times, when high-speed train technology needed them. So it is that planning for our own survival and prosperity, also requires a generous, caring long term view, to contribute to a changed world that our successors can enjoy.

These thoughts have started in the remote past, scanned through the ages to modern times, studied the different currents of the present, and finally, turned to the future. Once again, the cold windswept shingle beach comes to mind, with the surging current of history scouring along it, wave after wave. Once again, we are the pebbles, but are we as helpless as the pebbles of past generations? Do we understand the leaping tide better? Can we now make a difference, even changing its huge flow, for worse or for better? Can we rise above the morass of our own struggles, and see a wider view, even across the horizon to the great beyond?

In our own expanding age, the answers begin to change from 'no' to 'yes'. If we learn the lessons of our heritage, if we apply the new knowledge rising from science, philosophy and the major religions—bringing them closer together—then we start to find the power to change our own destinies. Even the smallest contribution will count, even the smallest acceptance of good over evil will change something for the better— even the tiniest scrap of paper in purse or wallet, with the bullet points of your future resolve—even that will alter humankind's prospects for the future by a dot. Join the dots together, and we build a new world.

That's our challenge—to embrace the chance to engage with the Western society that embraces every one of us, and to contribute to a global world our flair for building compassionate, caring situations in the places we live, or can contact. We, the ordinary people, now *can* make a difference to our world. At the end of our own personal journeys, we now *can* leave behind something which is better than the conditions we found at the start.

Take another look, a fresh one, at the analogy of the watermill[2] in Chapter Eight. Can we extract some of the power of creation, to make its continuation more beautiful?

An Alternative Last Paragraph

For the minority—the religious seekers after truth—I'd like to close with a different vocabulary. For them the excitement of new beginnings comes from discovering a new freedom to relate to the incomparable greatness and power of an all creating God. This grows in us, in ever new ways and thoughts, and draws us more deeply into involvement with the accelerating global world around us. For the Christians in society, their journey draws us into the 'love' of God, to take the parentheses off that misunderstood word, and to find a radical new understanding of it. To do this we need to develop a new vocabulary of belief, but using the same knowledge base and processes that inform the rest of society. The need is to dump the archaic, and to discover new meanings. It is to find a new understanding of biblical history, and to edge away the layers of later superimposed meanings. It is to use the radicalism which Christ showed to cast away restrictive visions of God and man, and to build a new vision which making sense to secular Late-Modern people, addressing their deepest fears of insecurity and intellectual loneliness.

This calls for a new reformation of conventional church life. It calls for a new renaissance which I start to glimpse, but cannot know if it will take root and succeed. But I think I can see what will happen if it does not.

The base of everything is to re-appraise the message of Jesus, and what he meant by the 'Kingdom of God'. For that calls us to journey forward to a transformation which makes us carry on our journeys in a different way. During the course of these, we know not when or how, the challenge is to die to a former way of being, and to rise to a new one where the presence of the creator God fills us, transforms us, and takes us past what we can now imagine or see. It was, after all, the ordinary people who went before us, who changed *their* world for us to inherit.[3]

NOTES TO CHAPTER TEN

1. Google *'the cosmic free lunch'* to introduce the debates between Stephen Hawking, David Birnbaum and others.

see also Peter Dickens and James Ormrod, *Cosmic Society: Towards a Society of the Universe* (Routledge, 2007).

2. *Watermill analogy.* See *pages 121-123.*

3. The original Christians of 1st century Thessalonica disturbed the local traditions and economy with their preaching, and were taken before the authorities as men who were 'turning the world upside down', a phrase from Acts 17:6 which reverberated through succeeding ages.

GLOSSARY

Abrahamic: The three related faith structures of Judaism, Christianity, and Islam.

Exponential: A rate of change that becomes quicker and quicker. In mathematics a graph of such a rate would not appear as a straight line, but as a curve that continually becomes steeper.

Homogenous: Strictly refers to people with the same ancestors. The similar word 'homogeneous' more broadly refers to groups of people of the same kind.

Isotropic: An object with properties that are the same when measured from different directions.

Palaeoanthropology: The study of the origins of the human species.

BIBLIOGRAPHY

SUGGESTIONS FOR FURTHER READING

BIBLES AND GOSPELS

Coogan, Michael D., et al, (ed) (2010) *The New Oxford Annotated Bible with the Apoccyrpha: NRSV, Fully revised 4th edition,* Oxford University Press.

Metzger, Bruce M. and Coogan, Michael D. (ed) (1993) *The Oxford Companion to the Bible,* Oxford University Press.

Miller, Robert J. (ed) (1994) *The Complete Gospels: Annotated Scholar's Version,* Polebridge Press.

Throckmorton, Burton H., Jr. (1992) *Gospel Parallels, NRSV 5th Edition: A Comparison of the Synoptic Gospels,* Thomas Nelson Publishers.

Zondervan Publishing (1984) *The Layman's Parallel Bible*

GREEK MYTHOLOGY

Graves, Robert (1955) *Greek Myths,* Penguin Books.

Kershaw, Stephen (2007) *Brief Guide to The Greek Myths,* Carroll & Graf Publishers.

DEAD SEA SCROLLS

Baigent, Michael and Leigh, Richard (1991) *The Dead Sea Scrolls Deception,* New York: Summit Books.

Eisenmann, Robert H., and Wise, Michael Owen (1992) *The Dead Sea Scrolls Uncovered: The First Complete Translation and Interpretation of 50 Key Documents Withheld for Over 35 Years,* Element Books.

HISTORICAL JESUS AND EARLY CHRISTIANITY

Borg, Marcus J., a range of works including:

—— (1994) *Meeting Jesus Again for the First Time: The Historical Jesus and the Heart of Contemporary Faith,* San Francisco: HarperCollins.

—— (1997) *The God We Never Knew: Beyond Dogmatic Religion to a More Authentic Contemporary Faith,* San Francisco: HarperCollins. (Named 'one of the ten best books in religion in 1997' *by Publishers Weekly.*)

—— (2001) *Reading the Bible Again for the First Time: Taking Bible Study Seriuosly but not Literally,* San Francisco: HarperCollins. (Included in *Publishers Weekly* 'ten best selling books in religion'.)

—— (2003) *The Heart of Christianity: Rediscovering a Life of Faith,* San Francisco: HarperCollins. (This has been used as a group study book in hundreds of churches.)

—— (2011) *Putting Away Childish Things: A Novel of Modern Faith,* HarperOne. (In this novel Borg presented his teachings on Christianity through fiction.)

with Crossan, John Dominic:

—— (2006) *The Last Week: What the Gospels Really Teach About Jesus' Final Days in Jerusalem,* New York: HarperCollins.

—— (2007) *The First Christmas: What the Gospels Really Teach Us About Jesus's Birth,* New York: HarperCollins.

—— (2009) *The First Paul: Reclaiming the Radical Visionary Behind the Church's Conservative Icon,* New York: HarperCollins.

Crossan, John Dominic—a range of works. The first three books listed below are considered to be the core:

——— (1991) *The Historical Jesus: The Life of a Mediterranean Jewish Peasant*, HarperCollins.

——— (1998) *The Birth of Christianity: Discovering What Happened in the Years Immediately After the Execution of Jesus*, HarperCollins.

——— with Reed, Jonathan L. (2004) *In Search of Paul: How Jesus's Apostle Opposed Rome's Empire with God's Kingdom*, HarperOne.

——— (2012) *The Power of Parable: How Fiction by Jesus Became Fiction about Jesus*, HarperCollins.

Funk, Robert Walter (1993) *The Five Gospels: What Did Jesus Really Say? the Search for the Authentic Words of Jesus*, Harper Collins.

Spong, John Shelby—a range of works including:

——— (1994) *Rescuing the Bible from Fundamentalism Resurrection—Myth or Reality*, HarperOne.

——— (1998) *Why Christianity Must Change or Die: A Bishop Speaks to Believers in Exile,* HarperOne.

——— (2001) *A New Christianity for a New World*, HarperOne.

——— (2009) *Eternal Life: A New Vision—Beyong Religion, Beyond Theism, Beyound Heaven and Hell*, HarperOne.

Theissen, Gerd and Mertz, Anette (1998) *The Historical Jesus—A Comprehensive Guide*, Fortress Press.

Theology and the Natural World

Johnson, Elizabeth (2011) *Ask the Beasts: Darwin and the God of Love*, Bloomsbury.

Modern Science

Cannato, Judy (2010) *Field of Compassion—How the New Cosmology is Transforming Spiritual Life*, Sorin Books.

McTaggart, Lynne (2008) *The Field: The Quest for the Secret Force of the Universe*, Harper Perennial.

Rovelli, Carlo (2015) *Seven Brief Lessons on Physics*, Allen Lane.

Sheldrake, Rupert (2009) *Morphic Resonance: The Nature of Formative Causation,* Park Street Press (4th, revised and expanded U.S. edition of *The New Science of Life: The Hypothesis of Morphic Resonance*).

Websites

Marcus J. Borg:
marcusjborg.com

Dominic Crossan:
www.johndominiccrossan.com

John Shelby Spong:
johnshelbyspong.com

Lynne McTaggart:
www.lynnemctaggart.com

David Birnbaum:
www.summametaphysica.com

Rupert Sheldrake:
www.sheldrake.org

Acknowledgements

I am grateful to the authors, editors and publishers listed in the Bibliography for their own works, without which this book could not have been written. I commend them all to readers with social consciences wanting to pursue these thoughts further. I also acknowledge the considerable practical support given to me by Lucy McCarraher and Rethink Press in producing this book. They have encouraged me at every turn and twist, and I commend them to any would-be author with their own story to tell.

I pay tribute to Peter Udell, John Driver, and Martin Camroux for their specialised advice. Mention of Martin's considerable support reminds me to acknowledge the help given by other clergy colleagues, including, Stephen Carter, David Knight and Don Macalister.

I also express my gratitude to Professor Saad Tahir, Professor Brian Davidson, Nigel Richardson Richard Laing and many other medical practitioners including Leone Rigden, Sue Rush and Liz Gardiner for keeping an old body running long enough to be able to pass these thoughts on to you, the readers!

THE AUTHOR

Chris Thorby has spent a life-time exploring the past of our civilisation and the complex ways in which we all link together today. Beginning with a BBC career of over thirty years in international broadcasting, he saw history created at first hand from the Cold War years, to the age of global communications and instant news on your mobile.

He watched this history unravel in this era. He saw how it changed people's prospects and future hopes, and the worlds they lived in. He also began to notice with a historian's eye, how many of the present problems were similar to those which past generations had managed to overcome. Could it be that experience of the past could better equip people today as they continue their current life journeys? Could it be that much of our new knowledge of science, and religion is actually converging, and providing us with new opportunities never known to humankind before? The book provides a guide to a wide range of modern learning, for non-specialists.

It offers readers practical ideas about how to take control of their journey and maximise the joys and value of making it. As he reveals a series of key points for our remaining time, Chris shows how the problems of our forebears are our problems too—and their successes in surviving them are there to guide us as well. There are indeed new ways to plan for the future, to build up personal strength and confidence by discovering where we have come from, where we are now and where we are going. His avid interest continued in a post-retirement career of voluntary work and leadership for the last twenty-five years, and he now writes this book to pass on what he has found.

Throughout these later years he and Annie his wife, brought up five children in a coastal village overlooking a wide tidal estuary, and as keen wildlife photographers have many pictures of waves crashing on the shingly beaches. As he thinks of modern global society, he wonders if these little pebbles are like people just being swept along by powerful sources they can't control. No, he says, no longer. Now we can start to control our individual futures.

The closing chapters show readers how to set about this, and *Life Journeys* ends with a final intriguing question about religion. Can it provide a series of 'extras' to overcome its own weaknesses, and the huge pressures of modern living? Like everything else in this guide to Journeying, it's for you to decide.